APPLIED ADHESIVE BONDING IN SCIENCE AND TECHNOLOGY

Edited by **Halil Özer**

Applied Adhesive Bonding in Science and Technology
http://dx.doi.org/10.5772/intechopen.68926
Edited by Halil Özer

Contributors

Yusuke Okabe, Gerald Schalau, Robert Huber, Alexis Bobenrieth, Linda Nartker, Xavier Thomas, Arif Nuryawan, Eka Mulya Alamsyah, Eva Malmström, Emelie Norström, Linda Fogelström, Farideh Khabbaz, Deniz Demircan, Sontipee Aimmanee, Elena Dinte, Bianca Sylvester, Halil Ozer

Published by InTech
Janeza Trdine 9, 51000 Rijeka, Croatia

Publishing Process Manager Maja Bozicevic
Technical Editor SPi Global
Cover InTech Design team

First published February, 2018
Printed in Croatia
Legal deposit, Croatia: National and University Library in Zagreb

Additional hard copies can be obtained from orders@intechopen.com

Applied Adhesive Bonding in Science and Technology, Edited by Halil Özer
p. cm.
Print ISBN 978-953-51-3839-6
Online ISBN 978-953-51-3840-2

For EU product safety concerns:
IN TECH d.o.o., Prolaz Marije Krucifikse Kozulić 3, 51000 Rijeka, Croatia,
info@intechopen.com or visit our website at intechopen.com.

Meet the editor

Halil Özer is currently a full professor at the Department of Mechanical Engineering of the Yıldız Technical University in Istanbul and the committee member of PhD proficiency in Mechanical Engineering. He completed his postdoctoral training at the Eindhoven University of Technology in the Netherlands, in the area of damage mechanics. He is the author and coauthor of scientific publications, especially in the field of the bi-adhesive joint, the self-healed adhesive joint, and cohesive zone modeling. He supervised PhD/MSc students mainly on adhesive joints.

Contents

Preface

Adhesives produced from the traditional tools have been used for thousands of years. Today, adhesives are used extensively in aerospace and industrial and medical applications. In addition to the development in adhesives, modern computers have also made it easier to evaluate the performance of the adhesive joints by giving the results in seconds with the help of numerical techniques.

The subject of the book is multidisciplinary in nature as it deals with adhesives drawn from the disciplines of chemical, mechanical, medical, biological, and other sciences. Advantages of using adhesive bonding can be summarized as follows: leading to a significant decrease in stress concentrations, providing more uniform stress distributions along the overlap length, providing savings in weight and cost, eliminating any cuts/holes in the joint, and so on. Adhesive bonding is often an appropriate choice for joining similar/dissimilar substrates (various substrate combinations, e.g., metal-to-metal, wood-to-wood, metal-to-composite, metal-to-wood, and rubber-to-skin).

This book is divided into three sections: "Structural Adhesive Bonding," "Wood Adhesive Bonding," and "Adhesive Bonding in Medical Applications." Each section presents the applications of the adhesive bonding in various different disciplines.

In the first section, some applications of structural bonding techniques are presented. It is known that stress concentrations developed at the bonded joints reduce their failure loads and hence joint strengths. In the first chapter, the role of the adhesive layer on stress distributions is discussed by using some applications of bi-adhesive and modulus-graded joints. In the second chapter, a unified mathematical model for predicting the joint stresses of the adhesive-bonded tubular-coupler joints under several types of load is formulated. In the third chapter, possible benefits of using STPE in the area of flexible electronics are studied to obtain a simple stretchable bonding system.

In the second section, the fourth chapter summarizes some of the most recent scientific literature regarding the development of green adhesives. It was reported that fundamental research is still required in order to have a strong bonding between the adhesive and substrates. In the fifth chapter, two types of isocyanate wood adhesives, commonly used in Indonesia, are studied.

The third section presents a comprehensive review on some medical applications of the adhesive bonding. The sixth chapter reviews the silicone-based adhesive technologies, applications, and characterization, emphasizing those self-adhesive materials often used in skin-contact applications including transdermal drug delivery and wound care device attachment. The last chapter presents a brief history of adhesive use. In addition, the chapter provides important review materials about the new generation of adhesives, based on modern technologies such as nanotechnology, polymers, and biomimetic adhesives.

This book brings together scientists and provides the reader with a comprehensive overview of some recent developments in the field of adhesive bonding with the contributions of internationally recognized authors. This book provides an important review on the adhesive bonding practices. I would like to express my gratitude to all the authors who contributed to this book. I hope that the book published in open access will help researchers to benefit from it.

Halil Özer
Yıldız Technical University
Faculty of Mechanical Engineering
Mechanical Engineering Department
Istanbul, Turkey

Section 1

Structural Adhesive Bonding

Chapter 1

Introductory Chapter: Structural Adhesive Bonded Joints

Halil Özer

Additional information is available at the end of the chapter

http://dx.doi.org/10.5772/intechopen.74229

1. Introduction

Adhesives produced from the traditional tools have been used for thousands of years. The advantages of using adhesively bonding techniques instead of classical mechanical fasteners can be listed as joining similar/dissimilar materials, significantly reducing the stress concentrations, providing more-uniform stress distributions along the overlap length, savings in weight and cost, eliminating any cuts/holes in the joint, etc. Adhesive bonding is often an appropriate choice for joining similar/dissimilar substrates (various substrate combinations, e.g., metal-to-metal, metal-to-composite, metal-to-rubber, metal-to-glass, metal-to-wood, etc.). Subject of adhesive bonding is also multidisciplinary in nature since it deals with adhesives drawn from the disciplines of chemical, mechanical, medical and medicine, biological, and other sciences. Adhesives have therefore become a key research area because of their potential applications. Today, adhesives are used extensively in aerospace, industrial, and medical applications. Three basic types of adhesively bonded joints used commonly are shown in **Figure 1**.

Choosing an appropriate joining technique is important to have strong joints. Single lap-joint (SLJ) is a simple joint type that allows for joining two adherends easily (**Figure 1a**). The slope of the scarf is the main factor determining the stresses developed on the inclined section of scarf joints (**Figure 1b**). Butt joint is another simple joining technique, and can have some disadvantages due to the small overlap area (**Figure 1c**). Each lap-joint type has therefore comparative advantages and disadvantages over the others. It is therefore important to choose the appropriate lap-joint type considering the application purposes. In addition to joint types, mechanical properties of adhesive/adherend materials, overlap length, thicknesses of adhesive/adherend, etc., affect stresses developed in the joint and hence the joint strength.

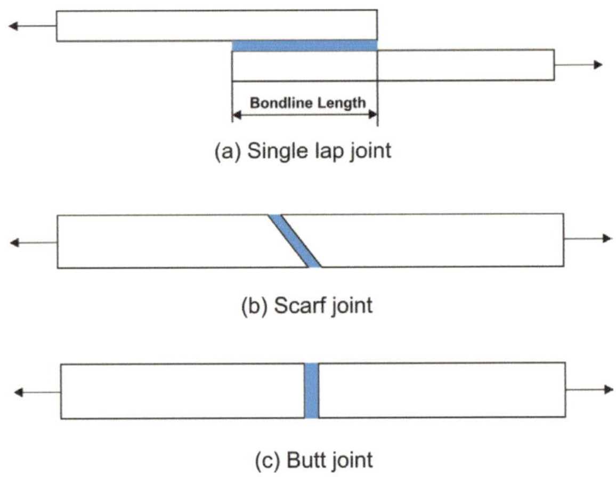

Figure 1. Three basic types of adhesively bonded joints.

This study aims to review some advancements in applications of the structural adhesively bonded joints. SLJ, the simplest form of adhesive joints, was introduced first and its characteristic behaviors of stress distributions along the bondline were given and discussed. In the adhesively bonded joint applications, reducing the stress concentrations and maximizing the failure load are the important issues to be solved. Many different techniques have been proposed to reduce stress concentrations. Grading the adhesive band has recently come to the fore among the reported remedies to overcome the problems faced by lap-joint applications. Two important applications of grading the adhesive band are using the bi-adhesive and modulus-graded bondlines in lap joints. The bi-adhesive bondline consists of three individual regions by a combination of stiff and flexible adhesives along the bondline, in which the flexible adhesives locate at the bondline ends and the stiff adhesive locates in the middle of the overlap. Second remedy is to use the modulus-graded bondline, in which bondline is graded functionally along the overlap length. This study aims to discuss the role of the adhesive layer on shear stress distributions and review some advancements in applications of the bi-adhesive and modulus-graded joints. Each joining technique was discussed briefly and compared with the joints bonded with a mono-adhesive alone.

2. Structural adhesive joints

2.1. Mono-adhesive bondline

Single lap joints have been studied by many authors [1–8]. As seen in **Figure 2**, load eccentricity results in developing the bending moments in an SLJ subjected to an axial loading. **Figure 2** shows the characteristic behavior of the peel and shear stress distributions along the bondline length.

It is seen from **Figure 2** that both shear and peel stresses become peak at overlap ends. However, higher peel stresses develop at the overlap ends due to bending moment effect. Single lap joints

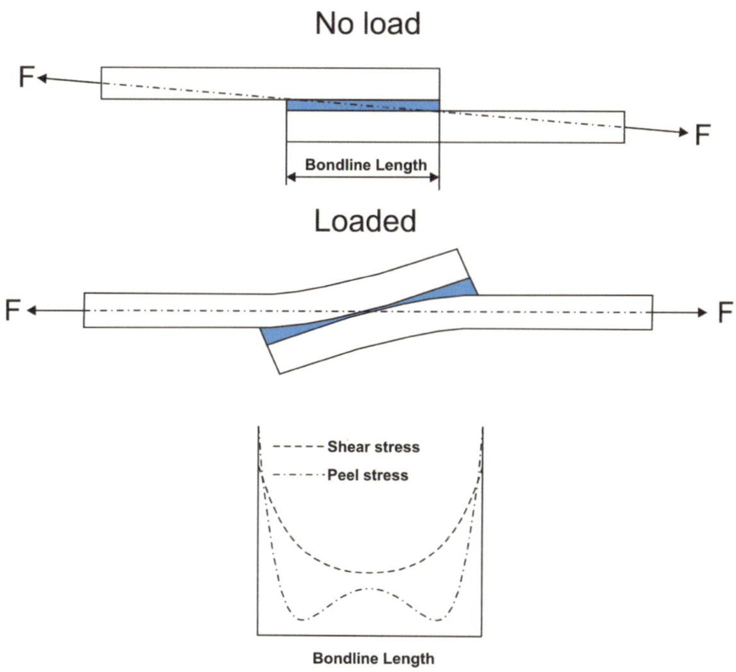

Figure 2. Deformed shape of an SLJ and stress distributions along its bondline length.

bonded with a mono-adhesive alone are widely used joints, however, weak in local stress concentrations at the overlap ends. High stress regions at the overlap ends reduce their failure loads and hence joint strengths. When adhesively bonded joint is subjected to tensile load, load is transferred mainly by shear stress developed in the adhesive layer. The first analytical model for assessing the adhesive shear stresses of bonded joints was developed by Volkersen [9]. His model is also called shear-lag model, which neglects the bending moment. The adhesive shear stress distribution is defined in the Volkersen model as follows:

$$\tau = \frac{F\omega}{2b}\frac{\cosh(\omega x)}{\sinh(\omega l/2)} + \left(\frac{t_t - t_b}{t_t + t_b}\right)\left(\frac{\omega l}{2}\right)\frac{\sinh(\omega x)}{\cosh(\omega l/2)} \tag{1}$$

where

$$\omega=\sqrt{\frac{G_a}{Et_t t_a}\left(1+\frac{t_t}{t_b}\right)} \tag{2}$$

In which, ω is the characteristic shear-lag distance. t_t and t_b are the top and bottom adherend thicknesses, respectively. *b* and *l* are the bond width and bond length, respectively. E, G_a, and F are the adherend modulus, adhesive shear modulus, and applied force, respectively. The x-axis passes through the mid-plane of the adhesive layer. As seen from **Figure 2**, high peel and shear stress concentrations develop at the overlap ends. There have been many attempts to reduce the stress concentrations of SLJ and improve its failure loads [10–12].

2.2. Modulus-graded bondline

One of the important remedies to overcome some deficiencies (i.e., stress concentration, decreasing in joint strength, etc.) arising in lap joints is grading the adhesive properties along the bondline. The earliest study on grading the modulus of an adhesive along the overlap length was performed by Raphael [14]. He splitted the adhesive bondline into finite number of discrete parts (**Figure 3**).

His model is based on the shear-lag concept of Volkersen, and therefore neglects the peel stress effect [13]. The work was undertaken as part of a program to design and test bonded rocket motor cases. The aim was to obtain the highest possible joint strength for a simple overlap. However, Raphael did not report any experimental work nor, indeed, quantify the possible benefits of a variable modulus bondline [14].

Figure 4 shows an SLJ bonded with the modulus-graded bondline. Recently, Carbas et al. [15] developed a simple analytical model to study the performance of the functionally graded joints.

In addition to numerical study, Carbas and others [16] also performed experimental study and used induction heating system to have a graded cure and joints with the adhesive gradually modified along the overlap. The induction system was set to allow the induction heating at the overlap ends and the induction cooling in the middle. They also performed analytical analyses to predict the failure load of the joints with graded cure and isothermal cure.

Modulus-graded joints have been studied in a limited number of papers in the literature and still open for numerical/analytical/experimental studies. The reader may refer to the following articles for current applications [17–23].

2.3. Bi-adhesive bondline

Bi-adhesive joint is an alternative stress-management technique for adhesively bonded lap joints. Its bondline includes a combination of stiff and flexible adhesives (**Figure 5**). This joint type including two types of adhesives in the overlap region is called as bi-adhesive, hybrid-adhesive, and mixed-adhesive joints in the literature.

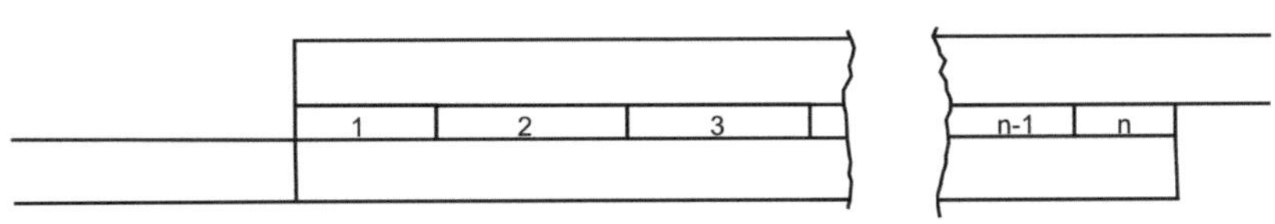

Figure 3. Raphael's modulus-graded bondline.

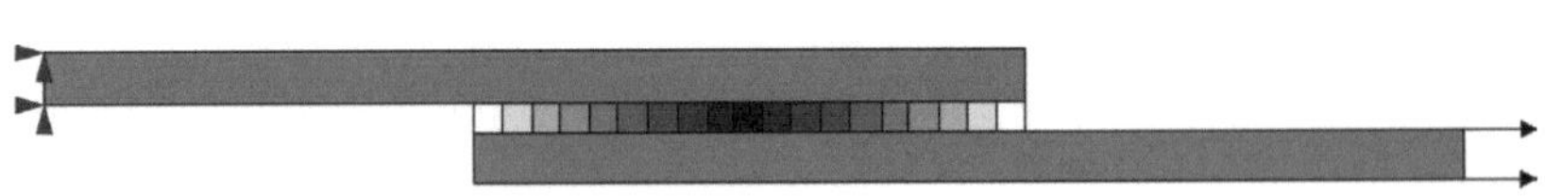

Figure 4. An SLJ bonded with the modulus-graded bondline.

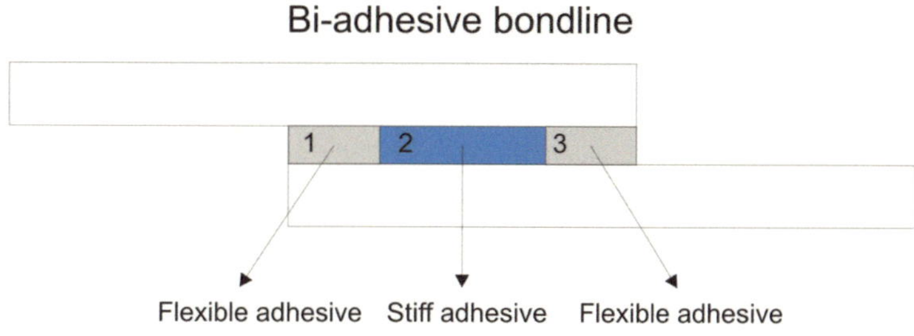

Figure 5. Bi-adhesive single lap-joint.

The stiff adhesive should be located in the middle and flexible adhesive at the ends of the bondline. The earliest study on the bi-adhesive joints was performed by Raphael [13]. Özer and Öz [24, 25] performed numerical studies to investigate the state of stress in the bi-adhesive bondline. In their other study [26], they also performed an experimental study to assess the effect of a bi-adhesive bondline on the failure load of both mono- and bi-adhesive SLJs. Their results were discussed briefly in **Figure 6**. **Figure 6** shows the characteristic behaviors of shear stress distributions along the mid-plane of the mono- and bi-adhesive layers. For comparison purposes, the shear stress distributions for the mono-flexible and mono-stiff adhesives were also given for mono-adhesive SLJs. As can be seen from **Figure 6**, shear stress distribution for the mono-flexible adhesive is more uniform than that of the mono-stiff adhesive and there is a lower stress concentration at the overlap edges. As a result, it is seen that the shear stress concentrations occurred at the overlap edges for mono-adhesive joints.

However, in the bi-adhesive bondline, as can be seen in **Figure 6**, the position of the maximum shear stress moves to a new position between adhesives (i.e., to the ends of the stiff adhesive in the middle). As reported above, the maximum shear stresses becomes peak at the overlap edges for mono-adhesive joints, however, it becomes peak at the contact interfaces for bi-adhesive joints.

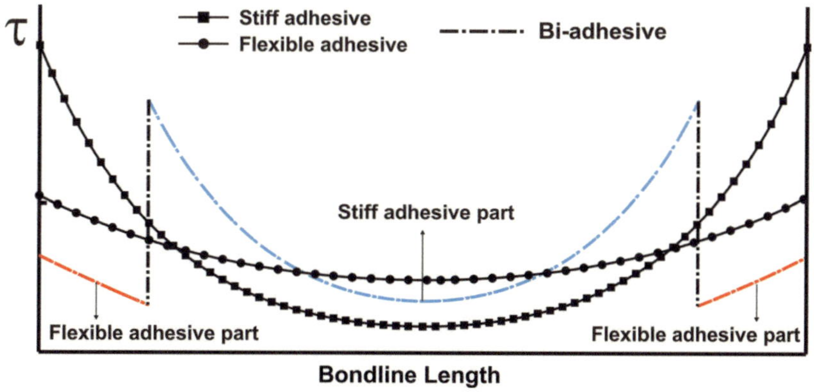

Figure 6. Characteristic behavior of shear stress distributions along mono- and bi-adhesive bondlines.

Therefore, it is seen that peak shear stress decreases at the overlap edges and increases at the contact interfaces (i.e., at the ends of the stiff adhesive) in the bi-adhesive bondline. It can be concluded that stiff adhesive in the middle contributes its high shear-strength-capacity to the bi-adhesive joint. Therefore, high stress concentrations at the bondline ends can be reduced by using bi-adhesive bondline. However, it is important to select the appropriate adhesive type for the bi-adhesive bondline. In addition, amounts of the stiff/flexible adhesives used in the bi-adhesive bondline also affect the shear stress values.

There are a limited number of publications in the open literature about the bi-adhesive joints. The reader may refer to the following articles for current applications of the joints bonded with bi-adhesive bondline [27–33].

3. Conclusion

It is known that high stress concentrations develop at the overlap ends of the adhesively bonded joints. Grading the adhesive band has recently come to the fore among the reported remedies to overcome the problems faced by lap-joint applications. In this study, the role of the adhesive layer on stress distributions was reviewed. Joining techniques using the bi-adhesive and modulus-graded bondlines were discussed briefly and compared with the joints bonded with a mono-adhesive alone. It is seen that high stress concentrations at the ends can be reduced by using these techniques. It is therefore concluded that stress concentration and joint strength can be optimized by using modulus-graded and bi-adhesive bondlines in the lap joints.

Author details

Halil Özer

Address all correspondence to: hozer@yildiz.edu.tr

Mechanical Engineering Department, Yıldız Technical University, Istanbul, Turkey

References

[1] Adams RD, Peppiatt NA. Effect of Poisson's ratio strains in adherends on stresses of an idealized lap joint. The Journal of Strain Analysis for Engineering Design. 1973;**8**:134-139

[2] Hart-Smith LJ. Adhesive-bonded single-lap joints. 1973. NASA CR-112236

[3] Adams RD, Peppiatt NA. Stress analysis of adhesive-bonded lap joints. The Journal of Strain Analysis for Engineering Design. 1974;**9**:185-196

[4] Allman DJ. A theory for elastic stresses in adhesive bonded lap joints. The Quarterly Journal of Mechanics and Applied Mathematics. 1977;**30**:415-436

[5] Bigwood DA, Crocombe AD. Elastic analysis and engineering design formulae for bonded joints. International Journal of Adhesion and Adhesives. 1989;**9**:229-242

[6] Tsai MY, Oplinger DW, Morton J. Improved theoretical solutions for adhesive lap joints. International Journal of Solids and Structures. 1998;**35**:1163-1185

[7] Zhao B, Lu Z-H. A two-dimensional approach of single-lap adhesive bonded joints. Mechanics of Advanced Materials and Structures. 2009;**16**:130-159

[8] Zhao B, Lu Z-H, Lu Y-N. Closed-form solutions for elastic stress–strain analysis in unbalanced adhesive single-lap joints considering adherend deformations and bond thickness. International Journal of Adhesion and Adhesives. 2011;**31**:434-445

[9] Volkersen O. Rivet strength distribution in tensile-stressed rivet joints with constant cross-section. Luftfahrorschung. 1938;**15**:41-47

[10] Hua Y, Gu L, Trogdon M. Three-dimensional modeling of carbon/epoxy to titanium single-lap joints with variable adhesive recess length. International Journal of Adhesion and Adhesives. 2012;**38**:25-30

[11] Belingardi G, Goglio L, Tarditi A. Investigating the effect of spew and chamfer size on the stresses in metal/plastics adhesive joints. International Journal of Adhesion and Adhesives. 2002;**22**:273-282

[12] Çalık A, Yıldırım S. Effect of adherend recessing on bi-adhesively bonded single-lap joints with spew fillet. Sadhana. 2017;**42**:317-325

[13] Raphael C. Variable-adhesive bonded joints. Journal of Applied Polymer Science: Applied Polymer Symposium. 1965;**3**:99-108

[14] Broughton JG, Fitton MD. Science of mixed-adhesive joints. In: da Silva L, Pirondi A, Öchsner A, editors. Hybrid Adhesive Joints. Berlin, Heidelberg: Springer; 2011. pp. 257-281

[15] Carbas RJC, da Silva LFM, Madureira ML, Critchlow GW. Modelling of functionally graded adhesive joints. The Journal of Adhesion. 2014;**90**:698-716

[16] Carbas RJC, da Silva LFM, Critchlow GW. Adhesively bonded functionally graded joints by induction heating. International Journal of Adhesion and Adhesives. 2014;**48**:110-118

[17] Stein N, Mardani H, Becker W. An efficient analysis model for functionally graded adhesive single lap joints. International Journal of Adhesion and Adhesives. 2016;**70**:117-125

[18] Stein N, Felger J, Becker W. Analytical models for functionally graded adhesive single lap joints: A comparative study. International Journal of Adhesion and Adhesives. 2017;**76**:70-82

[19] Nimje SV, Panigrahi SK. Strain energy release rate based damage analysis of functionally graded adhesively bonded tubular lap joint of laminated FRP composites. The Journal of Adhesion. 2017;**93**:389-411

[20] Stein N, Rosendahl PL, Becker W. Homogenization of mechanical and thermal stresses in functionally graded adhesive joints. Composites Part B: Engineering. 2017;**111**:279-293

[21] Khan MA, Kumar S. Interfacial stresses in single-side composite patch repairs with material tailored bondline. Mechanics of Advanced Materials and Structures. 2018;**25**:304-318

[22] Stapleton SE, Weimer J, Spengler J. Design of functionally graded joints using a polyurethane-based adhesive with varying amounts of acrylate. International Journal of Adhesion and Adhesives. 2017;**76**:38-46

[23] Kumar S, Wardle BL, Arif MF. Strength and performance enhancement of bonded joints by spatial tailoring of adhesive compliance via 3D printing. ACS Applied Materials and Interfaces. 2017;**9**:884-891

[24] Özer H, Öz Ö. Three dimensional finite element analysis of bi-adhesively bonded double lap joint. International Journal of Adhesion and Adhesives. 2012;**37**:50-55

[25] Özer H, Öz Ö. A comparative evaluation of numerical and analytical solutions to the biadhesive single-lap joint. Mathematical Problems in Engineering. 2014;**2014**:852872

[26] Öz Ö, Özer H. An experimental investigation on the failure loads of the mono and bi-adhesive joints. Journal of Adhesion Science and Technology. 2017;**31**:2251-2270

[27] das Neves PJC, da Silva LFM, Adams RD. Analysis of mixed adhesive bonded joints. Part I: Theoretical formulation. Journal of Adhesion Science and Technology. 2009;**23**:1-34

[28] Yousefsani SA, Tahani M. Relief of edge effects in bi-adhesive composite joints. Composites Part B: Engineering. 2017;**108**:153-163

[29] Breto R, Chiminelli A, Lizaranzu M, Rodríguez R. Study of the singular term in mixed adhesive joints. International Journal of Adhesion and Adhesives. 2017;**76**:11-16

[30] Temiz S. Application of bi-adhesive in double-strap joints subjected to bending moment. Journal of Adhesion Science and Technology. 2006;**20**:1547-1560

[31] Marques EAS, Campilho RDSG, da Silva LFM. Geometrical study of mixed adhesive joints for high-temperature applications. Journal of Adhesion Science and Technology. 2016;**30**:691-707

[32] Akpinar S, Aydin MD, Özel A. A study on 3-D stress distributions in the bi-adhesively bonded T-joints. Applied Mathematical Modelling. 2013;**37**:10220-10230

[33] Chiminelli A, Breto R, Izquierdo S, Bergamasco L, Duvivier E, Lizaranzu M. Analysis of mixed adhesive joints considering the compaction process. International Journal of Adhesion and Adhesives. 2017;**76**:3-10

Chapter 2

A Unified Analysis of Adhesive-Bonded Cylindrical Coupler Joints

Sontipee Aimmanee

Additional information is available at the end of the chapter

http://dx.doi.org/10.5772/intechopen.72288

Abstract

In the past years, many studies have been conducted on behaviors of adhesive tubular joints subjected to various loading conditions, such as torsion, axial, and internal and external pressure. However, the previous models are conceptually distinct, since they were developed to analyze only for each type of load. Mostly, homogeneous isotropic or orthotropic material were considered and thin-walled joint structures were examined. Therefore, the aim of this chapter is to present for the first time a generalized mathematical formulation and modeling of adhesive-bonded cylindrical coupler joints taking into account all loading scenarios. The inner and outer adherends can be made of isotropic, orthotropic, or laminated composite materials, and they are modeled as three-dimensional elastic body, so adherends with any thickness can be analyzed. Assumptions of an axisymmetric joint with linearly elastic adherends and adhesive materials are employed. Thin adhesive layer is considered so that only the out-of-plane adhesive stresses are concerned, and they are treated to be uniform through its thickness. Using elasticity theory and the newly developed finite-segmented method, stress distributions in both adherends and adhesive can be evaluated. Calculation examples of laminated composite joints are given. This model provides the unified analysis of adhesive-bonded cylindrical coupler joints.

Keywords: adhesive, coupler joint, lap joint, elasticity, finite-segment method

1. Introduction

Structures usually need to have joints connecting each part together due to the limitation of manufacturing, transportation, and installation. These structures are generally vulnerable at the joints because of the stress concentrations from material discontinuity. There are many types of joint, such as mechanical joints, welding joints, and adhesive-bonded joints. Over the other kinds of joints, adhesive-bonded joints have advantages due to less stress concentration, higher capability of joining dissimilar materials, lighter weight, and better corrosion resistance. Nonetheless, stress concentration in the adhesive layer is still existing. The adhesive stress distribution is locally nonuniform and always highest at the edges of the bonding region. Thus, in order to use an adhesive-bonded joint safely, it is important to predict the developed adhesive stress accurately. A good analysis also provides the understanding of the joint behavior, yielding a design for improving the joint performance by decreasing the joint stress concentration.

There are lots of literature dealing with the stresses in lap joints and coupler joints between two tubular adherends. Among several types of adhesive-bonded joint, cylindrical or tubular joints subjected to axial, torsional, and external as well as internal pressure loads have been one of the main focus of mechanics of adhesion research for a half century due to their popular usage in many engineering applications. The analytical modeling and finite-element analysis are the two popular approaches for predicting the stresses developed in adhesive and adherends. In order to recognize the progress in this field, some examples of important work in mathematical modeling are given below.

For axial loads, the early investigation was conducted by Lubkin and Reissner [1] who analyzed the tubular lap joints. Axial shear stresses and radial normal stresses in the adhesive layer were predicted under the assumption of approximating the layer as an infinite number of tensile and shear springs. However, disappearance of axial shear stress on the free surfaces at the ends of the adhesive was not considered. Adam and Peppiatt [2] performed axisymmetric finite-element analysis (FEA) of tubular lap joints subjected to stretching and twisting loads. The effects of an adhesive filet and partial tapering adherends on stress distribution were also reported. Using a minimum strain energy, Allman [3] proposed two-dimensional analytical solution for lap joints that ensure the traction-free boundary condition. Bending, stretching, and shearing of the adherends and shearing and tearing of the adhesive layer were taken into account. Shi and Cheng [4] formulated closed-form solutions for tubular lap joints utilizing the variational principle of complementary energy. Boundary conditions and assumptions of Allman were adopted in their model development. Nemes et al. [5] further developed the stress analysis of adhesive in a cylindrical assembly of two tubes. Variational method of the potential energy was also employed. Nonetheless, Nemes et al. neglected radial stress component in the joint. Kumar [6] presented a theoretical framework for the stress analysis of shaft-tube adhesive joints subjected to tensile loads. The joint assembly was considered to consist of similar or dissimilar isotropic or orthotropic adherends. The principle of minimum complementary energy and a stress function approach were used to establish the governing equations in order to determine the stress state in each constituent. To reduce the stress concentration, Kumar [7] also studied the use of functionally graded adhesive in a tubular lap joint with an isotropic adherend under tension. In his model, the adhesive was divided into annular rings to take into account the gradient property of shear modulus.

Regarding the case of the tubular joint subjected to torsion, Volkersen [8] provided a closed-form solution for circumferential shear stresses at the interface of tubular lap joint exerted by a torque. Pugno and Surace [9] investigated the analysis of the joint subjected to torsion. They utilized the common function of resultant torques in adherends and achieved the uniform adhesive hoop shear stress by tapering adherend surfaces. Xu and Li [10] investigated the full three-dimensional stress analysis of a bonded tubular-coupler joint subjected to torsion. Their purpose was to investigate all of the adherend and adhesive stress components without the assumption of through-thickness constant stresses across the adhesive layer. Oh [11] performed an analysis of the bonded tubular lap joint of laminated tubes with softening adhesive's stiffness properties under torsion using an elasticity model. Oh concluded that the load capacity in the linear analysis can be quite underestimated when compared to the nonlinear modeling. Spaggiari and Dragoni [12] investigated the joint studied in the Kumar's work in [7], but the joint is subjected to torsion instead. They developed the closed-form function of the adhesive shear modulus in order to minimize adhesive shear stress over the bonding region and addressed the limitation of shear modulus and thicknesses ratio for joint manufacturing with functionally graded adhesives. Recently, Aimmanee and Hongpimolmas [13] formulated a mathematical model of an adhesive-bonded tubular joint with a variable-stiffness composite coupler. The optimal variable fiber orientation in the coupler was determined to minimize the adhesive hoop shear stress.

Stresses in cylindrical joints under pressure have also been studied, even though investigation of this type of load was comparatively scarce compared to the above two loadings. Terekhova and Skoryi [14] provided a close-form solution for the stresses in tubular lap joints under external and internal pressures and axial forces. Their model neglected the effect of adherend bending. Baishya et al. [15] conducted research in individual and combined effect of internal pressure and torsional loading on stress and failure characteristics of tubular single lap joints made of composite materials. The onset of different joint fracture modes was investigated in their work. Strength analysis of adhesive joints of riser pipes in deep sea environment loadings was performed by Zhang et al. [16] External pressure, internal pressure, tension, torsion, and bending were examined to understand singular stress fields existing around end of the interface. Apalak [17] investigated elastic stresses in the adhesive layer and tubes of an adhesively bonded tubular joint with functionally graded tubes subjected to an internal pressure. Finite-element method was used to model the tubes having gradient layer between a ceramic layer and a metal layer.

According to the former analytical research work presented in the literature, the problems can be mathematically complicated even though the joint is made of simple conventional isotropic adherends. In addition, the previous models usually are distinct for each type of load, since they were developed to analyze only for a specific loading case. Therefore, this chapter aims to present a mathematical modeling of adhesive-bonded cylindrical coupler joints taking into account all loading scenarios, i.e., torsion, axial, and pressure loadings. The inner and outer adherends can be made of isotropic, orthotropic, or laminated composite materials, and they are modeled as three-dimensional elastic body, so a thick or solid cylinder adherend can also be analyzed. Stresses in adhesive layer and adherends can be evaluated by newly developed finite-segment method. The unified formulation of the model will be discussed in the next section.

2. Elasticity theory of a laminated cylindrical structure

A bonded-coupler joint is illustrated in **Figure 1(a)**. The joint consists of two inner tubes (adherend part 1), a coupler of length 2 L (or adherend part 2), and an adhesive layer. The cylindrical coordinates (x, θ, r) depicted in **Figure 2** are used to describe the joint geometry. Because of symmetry about the cross-sectional plane in the middle, only half of the coupler joint is demonstrated in **Figure 1(b)**. Either half is equivalent to a single tubular lap joint and applicable for modeling and analysis. The inner tubes are considered to be made of an ordinary material, such as an isotropic metal or a more sophisticated material, namely, orthotropic material or laminated composite. On the contrary the coupler is proposed to be fabricated from a symmetric-balanced laminated composite with variable fiber orientation in the x direction.

For the sake of generality, this section discusses the elasticity theory of a laminated cylindrical tube [18]. A sketch of a general open-ended, cylindrical, laminated N-layer tube subjected to uniform loads is shown in **Figure 2**. Each layer is made of a unidirectional fiber-reinforced composite material. The principal material coordinates (1, 2, 3), whose axes are mutually orthogonal, are defined along the fiber orientation, tangent, and normal to the tube surface, respectively. The layers in the tube are perfectly bonded between each other. Evidently, this considered laminated cylinder can be simply degenerated into a single isotropic or orthotropic tube by letting $N = 1$ and employing the related elastic properties. For the tube with

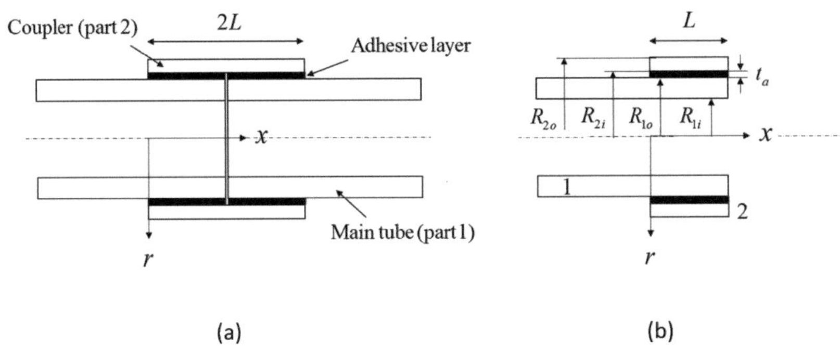

Figure 1. Schematic of a bonded-coupler joint: (a) full model and (b) half model or tubular lap joint model.

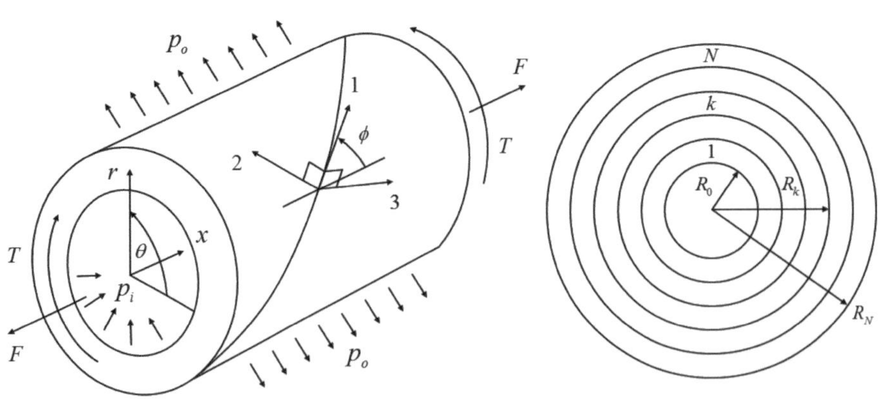

Figure 2. A laminated tube and the defined coordinate systems.

axisymmetric geometry and circumferentially independent material properties under a uniform load, the strain–displacement relations in the kth layer in the cylindrical coordinates are

$$\begin{aligned} \varepsilon_x^{(k)} &= \frac{\partial u^{(k)}}{\partial x}, & \varepsilon_\theta^{(k)} &= \frac{w^{(k)}}{r}, & \varepsilon_r^{(k)} &= \frac{\partial w^{(k)}}{\partial r} \\ \gamma_{\theta r}^{(k)} &= \frac{\partial v^{(k)}}{\partial r} - \frac{v^{(k)}}{r}, & \gamma_{xr}^{(k)} &= \frac{\partial u^{(k)}}{\partial r}, & \gamma_{x\theta}^{(k)} &= \frac{\partial v^{(k)}}{\partial x} \end{aligned} \tag{1}$$

where ε and γ denote normal and shear strains, respectively. u, v, and w are displacements in axial, tangential, and radial directions, respectively. Superscript (k) indicates that the corresponding quantities are in the kth layer.

According to the prescribed loading conditions and constant fiber orientation, the normal stresses, σ, and the shear stresses, τ, are independent of x and θ. The equilibrium equations in the kth layer along the r-, θ-, x-directions are reduced to ordinary differential equations with respect to r, respectively, as

$$\frac{\partial \sigma_r^{(k)}}{\partial r} + \frac{1}{r}\left(\sigma_r^{(k)} - \sigma_\theta^{(k)}\right) = 0 \tag{2}$$

$$\frac{\partial \tau_{\theta r}^{(k)}}{\partial r} + \frac{2}{r}\tau_{\theta r}^{(k)} = 0 \tag{3}$$

$$\frac{\partial \tau_{xr}^{(k)}}{\partial r} + \frac{1}{r}\tau_{xr}^{(k)} = 0 \tag{4}$$

The stresses and strains in the kth layer in (x, θ, r) coordinates expressed in Eqs. (1)–(4) can be transformed to those in the principal material coordinates (1, 2, 3) as follows:

$$\{\sigma\}_{123}^{(k)} = [T]^{(k)}\{\sigma\}_{x\theta r}^{(k)}, \qquad \{\varepsilon\}_{123}^{(k)} = [T]^{(k)}\{\varepsilon\}_{x\theta r}^{(k)} \tag{5}$$

where $\{\sigma\}_{123}^{(k)}$ and $\{\varepsilon\}_{123}^{(k)}$ are tensorial stress and tensorial strain components, respectively. $[T]^{(k)}$ is transformation matrix of the kth layer as shown in Eq. (6), in which $m^{(k)} = \cos Ø^{(k)}$ and $n^{(k)} = \sin Ø^{(k)}$. $Ø^{(k)}$ is fiber angle of the kth layer as shown in **Figure 2**.

$$[T]^{(k)} = \begin{bmatrix} \left(m^{(k)}\right)^2 & \left(n^{(k)}\right)^2 & 0 & 0 & 0 & 2m^{(k)}n^{(k)} \\ \left(n^{(k)}\right)^2 & \left(m^{(k)}\right)^2 & 0 & 0 & 0 & -2m^{(k)}n^{(k)} \\ 0 & 0 & 1 & 0 & 0 & 0 \\ 0 & 0 & 0 & m^{(k)} & -n^{(k)} & 0 \\ 0 & 0 & 0 & n^{(k)} & m^{(k)} & 0 \\ -m^{(k)}n^{(k)} & m^{(k)}n^{(k)} & 0 & 0 & 0 & \left(m^{(k)}\right)^2 - \left(n^{(k)}\right)^2 \end{bmatrix} \tag{6}$$

The constitutive relation in the kth layer in the cylindrical coordinates can be written as

$$\{\sigma\}_{x\theta r}^{(k)} = \left[\overline{C}\right]^{(k)}\{\varepsilon^{eng}\}_{x\theta r}^{(k)} \tag{7}$$

In the above, $\boxed{C}^{(k)}$ is the transformed stiffness matrix, and $\{\varepsilon^{eng}\}_{x\theta r}^{(k)}$ is engineering strain components in the global cylindrical coordinate system. The transformed stiffness matrix $\boxed{C}^{(k)}$ can be evaluated as

$$\boxed{C}^{(k)} = \left\{[T]^{(k)}\right\}^{-1}[C]^{(k)}[R][T]^{(k)}[R]^{-1} \tag{8}$$

where $[C]^{(k)}$, as shown in Eq. (9), is the stiffness matrix in the principle material coordinate system in the kth layer.

$$[C]^{(k)} = \begin{bmatrix} 1/E_1^{(k)} & -\upsilon_{12}^{(k)}/E_1^{(k)} & -\upsilon_{31}^{(k)}/E_3^{(k)} & 0 & 0 & 0 \\ -\upsilon_{21}^{(k)}/E_2^{(k)} & 1/E_2^{(k)} & -\upsilon_{32}^{(k)}/E_3^{(k)} & 0 & 0 & 0 \\ -\upsilon_{13}^{(k)}/E_1^{(k)} & -\upsilon_{23}^{(k)}/E_2^{(k)} & 1/E_3^{(k)} & 0 & 0 & 0 \\ 0 & 0 & 0 & 1/G_{23}^{(k)} & 0 & 0 \\ 0 & 0 & 0 & 0 & 1/G_{13}^{(k)} & 0 \\ 0 & 0 & 0 & 0 & 0 & 1/G_{12}^{(k)} \end{bmatrix}^{-1} \tag{9}$$

$E^{(k)}$ and $G^{(k)}$ are Young's modulus and shear modulus, respectively. $[R]$ is the Reuter's matrix, which is defined as

$$[R] = \begin{bmatrix} 1 & 0 & 0 & 0 & 0 & 0 \\ 0 & 1 & 0 & 0 & 0 & 0 \\ 0 & 0 & 1 & 0 & 0 & 0 \\ 0 & 0 & 0 & 2 & 0 & 0 \\ 0 & 0 & 0 & 0 & 2 & 0 \\ 0 & 0 & 0 & 0 & 0 & 2 \end{bmatrix} \tag{10}$$

With the strains defined in Eq. (1), three out of six equations of the compatibility in the cylindrical coordinates described in [19] are automatically satisfied. Solving the equilibrium equations in Eqs. (2)–(4) and using the strain–displacement relations in Eq. (1), the constitutive relation in Eq. (7), the remaining three compatibility equations, as well as the displacement continuity between each layer yield the displacement expressions in the kth layer of the laminated tube as illustrated in Eqs. (11)–(13):

$$u^{(k)}(x) = \varepsilon_x^0 x \tag{11}$$

$$v^{(k)}(x,r) = \gamma_{x\theta}^0 xr \tag{12}$$

$$w^{(k)}(r) = A_1^{(k)} r^{\lambda^{(k)}} + A_2^{(k)} r^{-\lambda^{(k)}} + \Gamma^{(k)}\varepsilon_x^0 r + \Omega^{(k)}\gamma_{x\theta}^0 r^2 \tag{13}$$

In the above, ε_x^0 and $\gamma_{x\theta}^0$ are axial strain and angle of twist per unit length constants, respectively. $A_1^{(k)}$ and $A_2^{(k)}$ are the integration constants in the kth layer. $\lambda^{(k)}$, $\Gamma^{(k)}$, and $\Omega^{(k)}$ are described in Eq. (14) in terms of components of the transformed stiffness matrix in the kth layer:

$$\lambda^{(k)} = \sqrt{\frac{\overline{C}_{22}^{(k)}}{\overline{C}_{33}^{(k)}}}, \quad \Gamma^{(k)} = \left(\frac{\overline{C}_{12}^{(k)} - \overline{C}_{13}^{(k)}}{\overline{C}_{33}^{(k)} - \overline{C}_{22}^{(k)}}\right), \quad \Omega^{(k)} = \left(\frac{\overline{C}_{26}^{(k)} - 2\overline{C}_{36}^{(k)}}{4\overline{C}_{33}^{(k)} - \overline{C}_{22}^{(k)}}\right) \tag{14}$$

Note that when a layer is made of 0° fiber orientation, the stiffness coefficients $\overline{C}_{16}^{(k)}$, $\overline{C}_{26}^{(k)}$, and $\overline{C}_{36}^{(k)}$ are zero and $\overline{C}_{12}^{(k)} = \overline{C}_{13}^{(k)}$ as well as $\overline{C}_{22}^{(k)} = \overline{C}_{33}^{(k)}$. As such, Eq. (13) is degenerated to become $w^{(k)}(r) = A_1^{(k)} r^{(k)} + A_2^{(k)} r^{-1(k)}$ [13].

For an N-layer laminated tube, there are 2 N + 2 unknown integration constants to be evaluated. Therefore, 2 N + 2 equations are required to solve for the constants. The first four equations written in Eqs. (15)–(18) are obtained from two equations of the force equilibrium with the external loads and two equations from the surface traction boundary conditions. Note that F in Eq. (15) is an axial force, T in Eq. (16) an applied torque, p_i in Eq. (17) a normal traction or internal pressure on the inner surface, and p_o in Eq. (18) a normal traction or external pressure on the outer surface. The remaining 2 N-2 equations can be obtained from N-1 continuity conditions of the interfacial radial stresses, σ_r, and N-1 continuity conditions of the interfacial radial displacements, w, as shown in Eqs. (19) and (20), respectively:

$$\int_{R_0}^{R_N} 2\pi\sigma_x r dr = 2\pi \sum_{k=1}^{N} \int_{R_{k-1}}^{R_k} \sigma_x^{(k)} r dr = F \tag{15}$$

$$\int_{R_0}^{R_N} 2\pi\tau_{x\theta} r^2 dr = 2\pi \sum_{k=1}^{N} \int_{R_{k-1}}^{R_k} \tau_{x\theta}^{(k)} r^2 dr = T \tag{16}$$

$$\sigma_r^{(1)}(R_0) = p_i \tag{17}$$

$$\sigma_r^{(N)}(R_N) = p_0 \tag{18}$$

$$\sigma_r^{(k)}(R_k) = \sigma_r^{(k+1)}(R_k) \qquad (k = 1, 2, 3, \ldots, N-1) \tag{19}$$

$$w^{(k)}(R_k) = w^{(k+1)}(R_k) \qquad (k = 1, 2, 3, \ldots, N-1) \tag{20}$$

Eqs. (15)–(20) give the system of algebraic equations written in matrix form as

$$\begin{bmatrix} k_{11} & k_{12} & k_{13} & \cdots & k_{1,2N+2} \\ k_{21} & k_{22} & k_{23} & \cdots & k_{2,2N+2} \\ k_{31} & k_{32} & k_{33} & \cdots & k_{3,2N+2} \\ k_{41} & k_{42} & k_{43} & \cdots & k_{4,2N+2} \\ \vdots & & & \ddots & \vdots \\ k_{2N+2,1} & k_{2N+2,2} & k_{2N+2,3} & \cdots & k_{2N+2,2N+2} \end{bmatrix} \begin{Bmatrix} F \\ T \\ p_i \\ p_o \\ 0 \\ 0 \\ \vdots \\ 0 \end{Bmatrix} = \begin{Bmatrix} \epsilon_x^0 \\ \gamma_{x\theta}^0 \\ A_1^{(1)} \\ A_2^{(1)} \\ \vdots \\ A_1^{(N)} \\ A_2^{(N)} \end{Bmatrix} \tag{21}$$

where k_{ij} $(i, j = 1, 2, 3, \ldots, 2N + 2)$ are the coefficients obtained from the equations above. By solving Eq. (21), the constants $\epsilon_x^0, \gamma_{x\theta}^0$, and $A_1^{(1)}, A_2^{(1)}, \ldots, A_1^{(N)}, A_2^{(N)}$ can be obtained. Subsequently, all displacements, strains, and stresses are calculated by using Eqs. (11)–(13), (1), and (7), respectively.

3. Formulation of an equivalent lap joint model

3.1. Derivation of governing equations

All geometric parameters of a perfectly bonded tubular lap joint are shown in **Figure 1(b)**. The adhesive is assumed to be isotropic and linearly elastic. The adhesive thickness t_a is considered to be very thin compared to the adherend thicknesses, and thus, the outer radius of part 1 R_{1o} is approximately the same as the inner radius of part 2, R_{2i}. In addition, there are only three out-of-plane stress components mainly contributed in the adhesive: hoop shear stress $\tau_{\theta r}^a$, longitudinal shear stress τ_{xr}^a, and radial normal stress σ_r^a. These stresses in the adhesive are treated to be uniform through the adhesive thickness. Applied torque T; applied axial force F; internal pressure exerted on the inner surface of adherend part 1, p_i or written specifically as p_{1i}; and external pressure exerted on the outer surface of adherend part 2, p_o or p_{2o}, are all included in the following formulation.

In order to derive the governing equations, let us initially consider the torque transmission through a coupler joint. The applied torque T is assumed to distribute only in the adherend part 1 and adherend part 2 as denoted as T_1 and T_2, respectively. Hence, the applied torque T can be written as

$$T = T_1 + T_2 \tag{22}$$

To determine the variation of the T_2 along the bonding length, the adherend 2 is divided into elements with an infinitesimal length dx. The equilibrium between the resultant torque in the element and the adhesive hoop shear stress can be expressed as follows:

$$\frac{1}{2\pi R_{2i}^2} \frac{dT_2(x)}{dx} = \tau_{\theta r}^a = G^a \gamma_{\theta r}^a \tag{23}$$

In Eq. (23), G^a is shear modulus of adhesive. By considering the deformation of an adhesive element on a cross-sectional plane in the overlap region of the perfectly bonded joint, the kinematic condition in the adhesive can be written as

$$\gamma_{\theta r}^a = \frac{v_{2i} - v_{1o}}{t_a} \tag{24}$$

and its derivative with respect to x is.

$$\frac{d\gamma_{\theta r}^a}{dx} = \frac{\gamma_{x\theta}^{2i} - \gamma_{x\theta}^{1o}}{t_a} \tag{25}$$

Combining Eqs. (23) and (25) yields the first governing equation:

$$\frac{d^2T_2(x)}{dx^2} = \frac{2\pi R_{2i}^2 G^a}{t_a}\left(\gamma_{x\theta}^{2i} - \gamma_{x\theta}^{1o}\right) \tag{26}$$

Next, consider equilibrium of resultant axial force. When the joint is subjected to tension or compression loads, the resultant axial force in the adherend 1, F_1, and in adherend 2, F_2, are produced at any given cross section in the overlap region, similar to Eq. (22). The force equilibrium is

$$F = F_1 + F_2 \tag{27}$$

The variation of the F_2 along the length can be examined by considering an infinitesimal elements in adherend part 2 with the differential length dx. The equilibrium between resultant axial force in the element and the adhesive longitudinal shear stress τ_{xr}^a can consequently be expressed as follows:

$$\frac{1}{2\pi R_{2i}}\frac{dF_2(x)}{dx} = \tau_{xr}^a = G^a\gamma_{xr}^a \tag{28}$$

By considering compatibility of the joint, it can be shown that.

$$\gamma_{xr}^a = \frac{u_{2i} - u_{1o}}{t_a} \text{ or } \frac{d\gamma_{xr}^a}{dx} = \frac{\varepsilon_x^{2i} - \varepsilon_x^{1o}}{t_a} \tag{29}$$

Combining Eqs. (28) and (29) yields the axial force governing equation:

$$\frac{d^2F_2(x)}{dx^2} = \frac{2\pi R_{2i}G^a}{t_a}\left(\varepsilon_x^{2i} - \varepsilon_x^{1o}\right) \tag{30}$$

Next, interacting through the adhesive thickness, the resultant normal traction acting on the outer surface of adherend 1, p_{1o}, and that exerting on the inner surface of adherend 2, p_{2i}, are generated. Under the assumption of thin adhesive layer, the resultant normal tractions p_{1o} and p_{2i} are related to each other as

$$p_{1o} = p_{2i} \tag{31}$$

Lastly, instead of directly equating adhesive radial normal stress to normal traction in (31), σ_r^a can be more accurately determined by the equilibrium equation in cylindrical coordinates of the adhesive layer showing in Eq. (32):

$$\frac{\partial\sigma_r}{\partial r} + \frac{1}{r}(\sigma_r - \sigma_\theta) + \frac{1}{r}\frac{\partial\tau_{\theta r}}{\partial\theta} + \frac{\partial\tau_{xr}}{\partial x} = 0 \tag{32}$$

With the conditions of axisymmetry, the equilibrium equation is reduced to

$$\frac{1}{R_{2i}}\left(\sigma_r^a - \sigma_\theta^a\right) + \frac{\partial \tau_{xr}^a}{\partial x} = 0 \tag{33}$$

According to the study conducted in [6], σ_θ^a is observed to have the same distribution as σ_r^a so they are legitimately regarded as being proportional to each other via adhesive normal stress ratio α. Their relation can be mathematically expressed in Eq. (34):

$$\sigma_\theta^a = \alpha \sigma_r^a \tag{34}$$

As a consequence, the equilibrium equation in Eq. (32) can then be written as

$$\sigma_r^a = -\frac{1}{2\pi(1-\alpha)}\frac{d^2F_2(x)}{dx^2} \tag{35}$$

3.2. Implementation of elasticity theory for adherends

The two governing equations Eqs. (26) and (30) have already been formulated to determine resultant loads in adherend part 2 of an adhesive-bonded-coupler joint. The resultant loads in adherend part 1 can be then calculated easily by using Eqs. (22) and (27) after all internal loads in adherend part 2 are evaluated. However, related through Eq. (21), the two equations are coupled and need to be solved altogether. To aptly deal with this complicated condition, the problem is separated into primary and secondary effects. When the joint is subjected to torsion, the hoop shear stress in the adhesive $\tau_{\theta r}^a$ is primary and dominant compared to the other adhesive stresses as discussed in [8, 10, 13], whereas in the case of the joint being under an application of longitudinal force, or external and internal pressure, the adhesive longitudinal shear stress τ_{xr}^a and adhesive radial normal stress σ_r^a are comparatively crucial [20]. By neglecting the secondary stress components and the corresponding resultant internal loads in the early calculation stage, the problem is then uncoupled and can be readily solved for the primary variables. The initially excluded stress components are later recovered by using the obtained solutions in the coupled set of governing equations.

First, further modification of the torque governing equation of Eq. (26) is performed, adherend in-plane shear strains $\gamma_{x\theta}^{1o}$ and $\gamma_{x\theta}^{2i}$ must be expanded in terms of the internal resultant loads. It can be seen that they are equal to $\gamma_{x\theta}^{01}R_{1o}$ and $\gamma_{x\theta}^{02}R_{2i}$, respectively, where $\gamma_{x\theta}^{01}$ and $\gamma_{x\theta}^{02}$ are denoted for $\gamma_{x\theta}^{0}$ of adherend parts 1 and 2. Utilizing Eq. (21) yields the relations:

$$R_{1o}\left(k_{21}^1F_1 + k_{22}^1T_1 + k_{23}^1p_{1i} + k_{24}^1p_{1o}\right) = \gamma_{x\theta,F}^{1o}F_1 + \gamma_{x\theta,T}^{1o}T_1 + \gamma_{x\theta,pi}^{1o}p_{1i} + \gamma_{x\theta,po}^{1o}p_{1o} = \gamma_{x\theta}^{1o} \tag{36}$$

$$R_{2i}\left(k_{21}^2F_2 + k_{22}^2T_2 + k_{23}^2p_{2i} + k_{24}^2p_{2o}\right) = \gamma_{x\theta,F}^{2i}F_2 + \gamma_{x\theta,T}^{2i}T_2 + \gamma_{x\theta,pi}^{2i}p_{2i} + \gamma_{x\theta,po}^{2i}p_{2o} = \gamma_{x\theta}^{2i} \tag{37}$$

where quantities k_{2X}^1 and k_{2X}^2, where $X = 1, 2, 3$, and 4 are the first four elements compliances in the second row of matrix $[k]$ in Eq. (21). Superscripts 1 and 2 are defined for adherend parts 1 and 2, respectively. $\gamma_{x\theta,F}^{1o}$, $\gamma_{x\theta,T}^{1o}$, $\gamma_{x\theta,pi}^{1o}$, $\gamma_{x\theta,po}^{1o}$, $\gamma_{x\theta,F}^{2i}$, $\gamma_{x\theta,T}^{2i}$, $\gamma_{x\theta,pi}^{2i}$, and $\gamma_{x\theta,po}^{2i}$ are in-plane shear strains per unit load on the outer interfacial surface of adherend part 1 and inner interfacial surface of adherend part 2, respectively.

By substituting Eqs. (36) and (37) into the governing equation Eq. (26) and utilizing Eqs. (22), (27), and (31), the governing equation in term of T_2 becomes

$$\frac{d^2T_2(x)}{dx^2} = K_F F_2(x) + K_T T_2(x) + K_p p_{2i}(x) + K_C \tag{38}$$

where the parameters K_F, K_T, K_p, and K_C are.

$$\begin{gathered} K_F = \frac{2\pi R_{2i}^2 G^a}{t_a}\left(\gamma_{x\theta,F}^{2i} + \gamma_{x\theta,F}^{1o}\right), K_T = \frac{2\pi R_{2i}^2 G^a}{t_a}\left(\gamma_{x\theta,T}^{2i} + \gamma_{x\theta,T}^{1o}\right), \\ K_p = \frac{2\pi R_{2i}^2 G^a}{t_a}\left(\gamma_{x\theta,pi}^{2i} - \gamma_{x\theta,po}^{1o}\right), \qquad K_C = \frac{2\pi R_{2i}^2 G^a}{t_a}\left(\gamma_{x\theta,po}^{2i} p_o - \gamma_{x\theta,pi}^{1o} p_i - \gamma_{x\theta,F}^{1o} F - \gamma_{x\theta,T}^{1o} T\right) \end{gathered} \tag{39}$$

Accompanying with the boundary conditions of Eq. (40), which are implied that torque in adherend part 2 is zero at $x = 0$ and fully transmitted at $x = L$, Eq. (38) is well-defined for solving resultant torque in the adherend part 2,T_2:

$$T_2(0) = 0, \qquad T_2(L) = T \tag{40}$$

Second, analogous to Eqs. (36) and (37), adherend in-plane normal strains ε_x^{1o} and ε_x^{2i} must also be written in terms of the internal resultant loads. Again, using the expression in Eq. (21), the following expressions are obtained:

$$(k_{11}^1)F_1 + (k_{12}^1)T_1 + (k_{13}^1)p_{1i} + (k_{14}^1)p_{1o} = \varepsilon_{x,F}^{1o}F_1 + \varepsilon_{x,T}^{1o}T_1 + \varepsilon_{x,pi}^{1o}p_{1i} + \varepsilon_{x,po}^{1o}p_{1o} = \varepsilon_x^{1o} \tag{41}$$

$$(k_{11}^2)F_2 + (k_{12}^2)T_2 + (k_{13}^2)p_{2i} + (k_{14}^2)p_{2o} = \varepsilon_{x,F}^{2i}F_2 + \varepsilon_{x,T}^{2i}T_2 + \varepsilon_{x,pi}^{2i}p_{2i} + \varepsilon_{x,po}^{2i}p_{1o} = \varepsilon_x^{2i} \tag{42}$$

In the above, k_{1X}^1 and k_{1X}^2, where $X = 1, 2, 3,$ and 4 are the first four elements in the first row of matrix $[k]$ in Eq. (21). Superscripts 1 and 2 are defined for adherend parts 1 and 2, respectively. $\varepsilon_{x,F}^{1o}$, $\varepsilon_{x,T}^{1o}$, $\varepsilon_{x,pi}^{1o}$, $\varepsilon_{x,po}^{1o}$ $\varepsilon_{x,F}^{2i}$, $\varepsilon_{x,T}^{2i}$, $\varepsilon_{x,pi}^{2i}$, and $\varepsilon_{x,po}^{2i}$ are newly denoted to indicate the physical meaning of the parameters. They represent the in-plane normal strains due to unit load on the adhesive-interfacial surface in the adherends. The unit load quantities are distinguished by after-comma subscripts F, T, or p. Quantities ε_x^{1o} and ε_x^{2i} are ε_x^0 of adherends 1 and 2, respectively.

Combining the governing equations Eqs. (30), (41), and (42), as well as the load equilibriums in Eqs. (22), (27), and (31), yields a new form of the axial force governing equation:

$$\frac{d^2F_2(x)}{dx^2} = k_F F_2(x) + k_T T_2(x) + k_p p_{2i}(x) + k_C \tag{43}$$

where the parameters k_F, k_T, k_p, and k_C are.

$$\begin{gathered} k_F = \frac{2\pi R_{2i} G^a}{t_a}\left(\varepsilon_{x,F}^{2i} + \varepsilon_{x,F}^{1o}\right), k_T = \frac{2\pi R_{2i} G^a}{t_a}\left(\varepsilon_{x,T}^{2i} + \varepsilon_{x,T}^{1o}\right) \\ k_p = \frac{2\pi R_{2i} G^a}{t_a}\left(\varepsilon_{x,\mathrm{pi}}^{2i} - \varepsilon_{x,po}^{1o}\right), \qquad k_C = \frac{2\pi R_{2i} G^a}{t_a}\left(\varepsilon_{x,po}^{2i} p_o - \varepsilon_{x,pi}^{1o} p_i - \varepsilon_{x,F}^{1o} F - \varepsilon_{x,T}^{1o} T\right) \end{gathered} \tag{44}$$

To specify the boundary conditions of Eq. (43), one can consider the disappearance of F_2 at $x = 0$. This is because the left end surfaces of the adherend are normal traction-free. The right end at $x = L$ on the other hand must take the full axial load F if there exists the application of external axial load. Thus, in the mathematical form, these boundary conditions are as follows:

$$F_2(0) = 0, \qquad F_2(L) = F \tag{45}$$

Finally, it should be noted that the occurrence of p_{2i} in Eq. (43) is closely related with the existence of F_2 because the tensile or compressive loading can induce the peeling traction. Therefore, p_{2i} in the equation is considered as unknown. However, it is possible to find the approximated relation between the two variables by letting T_2 and p_{2i} be zero; F_2 then can be evaluated and expressed in Eq. (46):

$$F_2(x) = a_0 e^{\sqrt{k_F}x} + b_0 e^{-\sqrt{k_F}x} - \frac{k_C}{k_F} \tag{46}$$

where a_0 and b_0 are integration constants.

Reinstating the resultant normal traction p_{2i} and substituting Eq. (46) into Eq. (43), it is found that p_{2i} can be simply estimated as

$$p_{2i} \approx a_1 e^{\sqrt{k_F}x} + b_1 e^{-\sqrt{k_F}x} \tag{47}$$

in which, a_1 and b_1 are unknown parameters. In order to determine these two parameters, two more boundary conditions are required from zero longitudinal shear stress τ^a_{xr} in the adhesive layer at left and right ends as shown in Eq. (48):

$$\left.\frac{dF_2}{dx}\right|_{x=0} = 0, \qquad \left.\frac{dF_2}{dx}\right|_{x=L} = 0 \tag{48}$$

Up to this point, the unified formulation of an analysis of adhesive-bonded coupler joint has been developed. The model can be universally used to determine the stresses in the adhesive layer for any particular load case previously mentioned. To elaborate the applicability of the model for each loading condition, i.e., torsion, axial, or external and internal pressure, the pertinent details are given below:

- For torsional load, the secondary variables, namely, F_2 and p_{2i}, are initially neglected in Eq. (38). In addition, $F = p_{1i} = p_{2o} = 0$. Consequently, T_2 and $\tau^a_{\theta r}$ can be evaluated. Subsequently, F_2, τ^a_{xr}, and σ^a_r and can be recovered and computed by employing the full form of Eqs. (43) and (47), (28), and (35), respectively.

- For axial load, the secondary variable T_2 is initially neglected. Additionally, $T = p_{1i} = p_{2o} = 0$. The primary variables F_2, τ^a_{xr}, and σ^a_r are solved by using the governing equation, Eqs. (43), (47), (28), and (35). T_2 and $\tau^a_{\theta r}$ is later calculated from the full form of Eqs. (38) and (23), respectively.

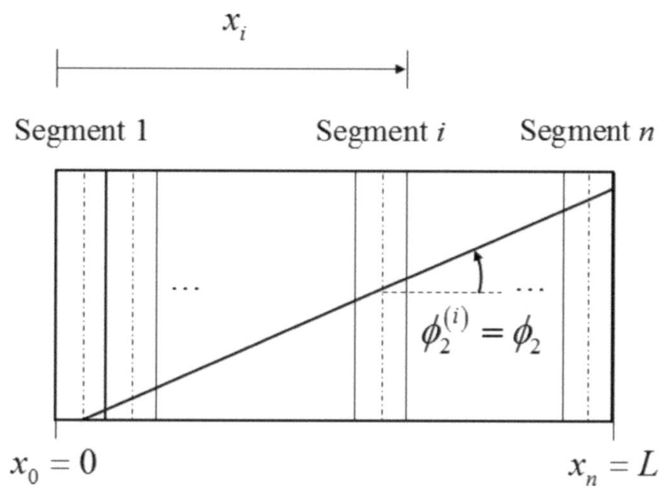

Figure 3. A segment at a certain axial position x.

- For external and internal pressure, the secondary variable T_2 is firstly omitted. In this case $T = F = 0$. If only external pressure is present, $p_{1i} = 0$, whereas if only internal pressure exists, $p_{2o} = 0$. The first variables F_2, τ_{xr}^a, and σ_r^a are solved by using the governing equation, Eqs. (43), (47), (28), and (35). T_2 and $\tau_{\theta r}^a$ can be later recovered the same way as those for the axial load.

3.3. Finite segment solution for evaluating adherend stresses

As previously discussed, all resultant loads in adherend parts 1 and 2 can be obtained as functions of x-coordinate. Equipped with the elasticity solution discussed in Section 2, the stress analysis in the adherends can be performed by utilizing a technique so-called finite-segment method (FSM) developed in [13]. The joint in the overlap region is divided into n numbers of segment as illustrated in **Figure 3** to take care of the axial variation of the resultant loads. The corresponding resultant loads of each segment are then approximated to be constant, and thus, application of the theory delineated in the previous section is valid. For a composite coupler, the fiber angle in the adherend part 2 is denoted as $\varnothing_2$. Note that if the stresses in adherend part 2 are evaluated, the internal pressure p_i in Eq. (21) or equivalently p_{2i} should be substituted by the radial normal stress σ_r^a obtained from Eq. (35) for more accurate results. Likewise, the procedure must be done for p_{1o} if adherend part 1 is considered. When all of the resultant loads in Eq. (21) in each segment are completely determined, stresses in both adherends can be finally computed and analyzed.

4. Results

Presented in this section are some sample computational results of the model developed. The numerical calculation is performed by using software MATHEMATICA™. The validation of the model is not given herein, since it has already been shown in [13] and [20]. Adhesive-bonded tubular joints with isotropic inner adherend and symmetric-balanced four-layer stacking sequence $[\pm Ø_2]_s$ couplers are selected for consideration. The reference joint geometry is given with parameters $R_{10} = 10$, $t_1 = 5$, $t_2 = 5$, $t_a = 0.1$, and $L = 40$ mm. The adherend part 1 is made of steel, whereas the adherend part 2 is fabricated from carbon fiber-reinforced plastic. Epoxy is used as the adhesive material. The material properties of the joints are listed in **Table 1**. Adhesive normal stress ratio, α, is set to be 20 because it has shown to provide accurate predictions of adhesive radial stresses [20]. In the following computational results, the adhesive stresses in the coupler joint are normalized by the average applied stress in each loading case, because the dimensionless stresses are readily exploited to identify the level of load distribution intensity in the joint.

4.1. Torsional loading

In the case of torsional loading, the joints are assumed to have a torque of 1 N.m as an input without loss of generality. Also, the adhesive mean shear stress τ_m^a in Eq. (49) is utilized to normalize the induced adhesive hoop shear stress in the coupler joint:

$$\tau_m^a = \frac{T}{2\pi R_{2i}^2 L} \tag{49}$$

The resultant torque of the adherend part 2 and normalized adhesive hoop shear stress can be calculated and plotted in **Figures 4** and **5**, respectively. It can be noticed that the joints considered develop the nonconstant slopes in **Figure 4** with relatively high torque gradients at both ends. This is equivalent to the peak adhesive hoop shear stresses at $x = 0$ and L in **Figure 5**. Note that the torque and stress distributions for $Ø_2 = 0^\circ$ and 30° are identical to those for 90° and 60°, respectively, so they cannot be clearly seen. In addition, since the fiber orientation of 45° is the most suitable angle to withstand the in-plane shear loads, the coupler with $Ø_2 = 45^\circ$ provides the lowest magnitude of $\tau_{\theta r}^a$ as expected.

Stress distributions in the composite coupler are illustrated in **Figure 6** for the case of $Ø_2 = 30^\circ$. The normal stress in the fiber direction σ_{11} in **Figure 6(a)** is the dominant stress component compared to those in the other directions. The radial normal stress σ_{33} illustrated in **Figure 6(b)** is relatively small at $x = 0$ mm and noticeably larger at $x = 40$ mm. In addition, σ_{33} at adhesive-

Properties	Epoxy (adhesive)	Steel (adherend 1)	Carbon/epoxy (adherend 2)
E_1 (GPa)	1.30	200.00	128.00
E_2, E_3 (GPa)	1.30	200.00	10.00
G_{12}, G_{13} (GPa)	0.46	76.90	50.00
G_{23} (GPa)	0.46	76.90	50.00
υ_{12}, υ_{13}	0.41	0.30	0.28
υ_{23}	0.41	0.30	0.47

Table 1. Mechanical properties of materials in the principle material coordinate [20].

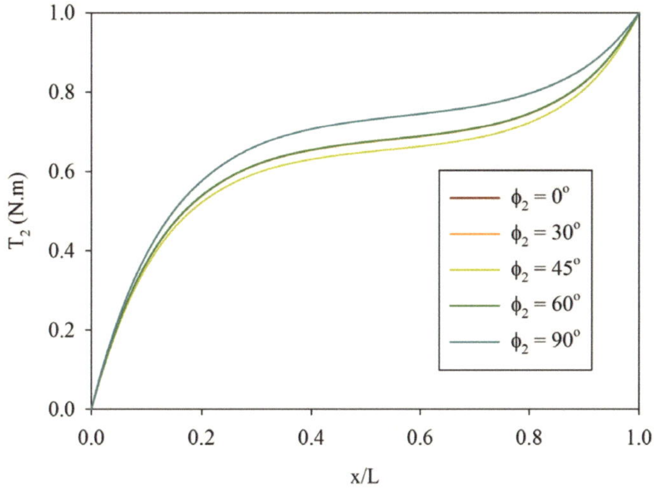

Figure 4. Torque distribution in adherend part 2 of composite coupler joint subjected to torsion.

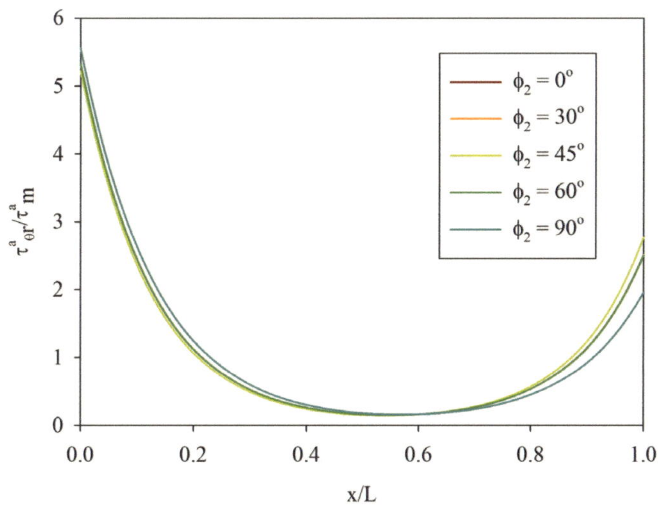

Figure 5. Adhesive hoop shear stress distribution of composite coupler joint subjected to torsion.

coupler interface r = 10.1 mm or σ_r^a is also minimal because of being the secondary effect. Note that the number of segments used to calculate the stresses in **Figure 6** is 40. Finally, the figure shows that the developed model is capable of capturing the variation of these two stress components through the coupler thickness thanks to the advantage of the elasticity theory.

4.2. Axial loading

When the coupler joints are subjected to an axial loading, a tension force with the magnitude of 1 N is used in calculation. For this particular case, the adhesive mean shear stress τ_m^a in Eq. (50)

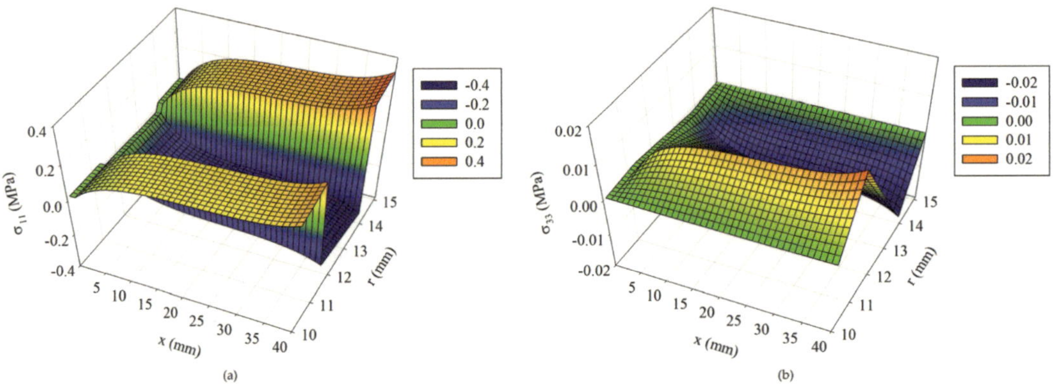

Figure 6. Distribution of stresses in composite coupler subjected to torsion.

is adopted to normalize the induced longitudinal shear stress and radial normal stress in the adhesive. Same as above, the normalized stresses can be utilized to indicate the distribution intensity of load transfer within the joints:

$$\tau_m^a = \frac{F}{2\pi R_{2i} L} \tag{50}$$

Figures 7 and **8** show the effect of fiber orientation on the distributions of F_2 and τ_{xr}^a along the overlap region, respectively. Observation in **Figure 8** reveals that by adjusting fiber orientation, the composite coupler can generate mostly uniform load transmission in the central bonding region. The internal forces F_2 of **Figure 7** in that region concomitantly reveal linear relationships with the spatial coordinate x/L. The optimum fiber angle $\varnothing_2$ is about 30°, which provides the lowest maximum τ_{xr}^a/τ_m^a of 1.2. **Figure 9** shows the radial normal stress in the adhesive. Generally speaking, the smaller peak of τ_{xr}^a/τ_m^a, the lower magnitude of σ_r^a/τ_m^a.

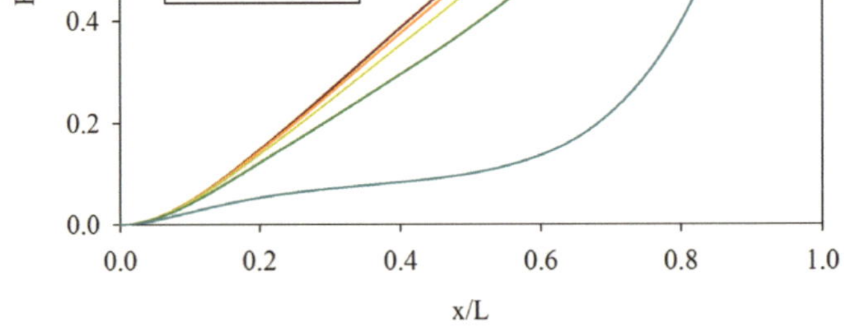

Figure 7. Force distribution in adherend part 2 of composite coupler joint subjected to tension.

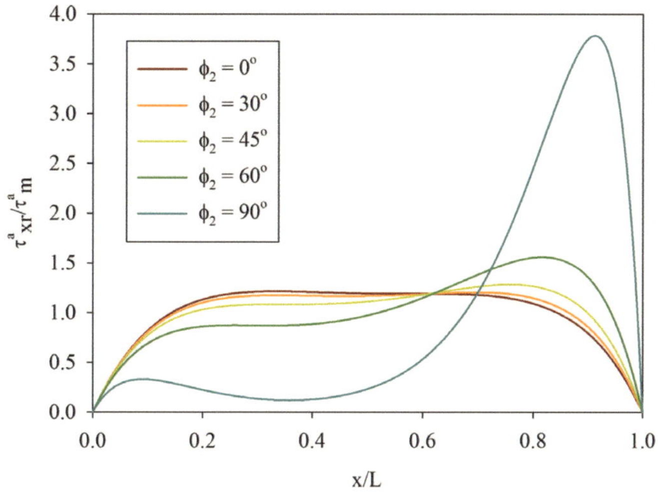

Figure 8. Adhesive axial shear stress distribution of composite coupler joint subjected to tension.

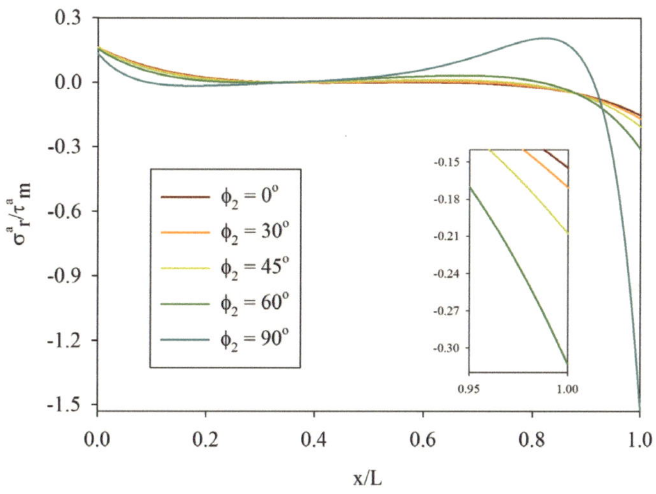

Figure 9. Interfacial radial stress distribution of composite coupler joint subjected to tension.

Figure 10 shows the normal stresses σ_{11} and σ_{33} of adherend part 2 in the principal material coordinate system, when the fiber orientation is equal to 30°. It can be seen from the figure that under the application of the axial force, σ_{11} is vanished at the left of the bonding region due to the traction-free surface, while σ_{33} is disappeared on the outermost area of the coupler. The stress in the fiber direction σ_{11} steadily attains the same maximum value along the bond length in all laminae of composite coupler. The radial normal stress σ_{33}, which is induced from the resultant axial force, is highest at the adhesive-coupler interface.

4.3. Pressure loading

Lastly, for the case of pressure loads, 1 MPa internal pressure is exerted inside the adherend part 1, but no external pressure is present on the outer surface of the adherend part 2. The

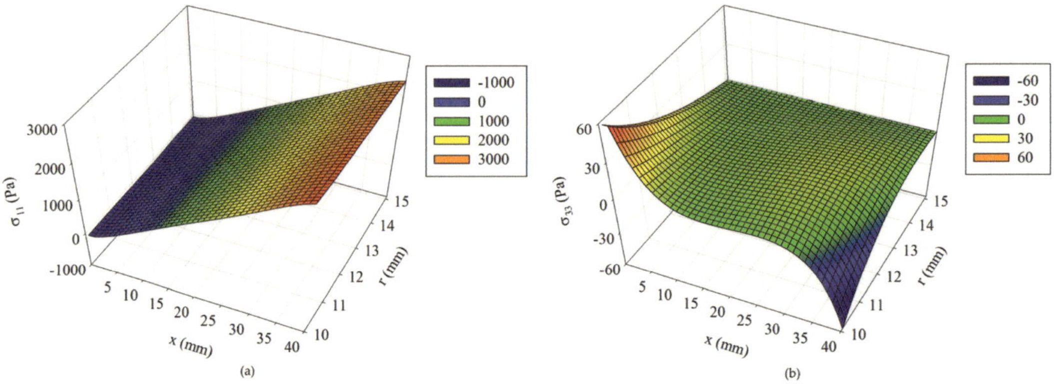

Figure 10. Distribution of stresses in composite coupler subjected to tension.

adhesive longitudinal shear stress and adhesive radial normal stress can then be normalized by the internal pressure to form the dimensionless variables.

Figures 11 and **12** show the effect of fiber orientation on the distributions of F_2 and τ_{xr}^a along the overlap region, respectively. **Figure 11** indicates that peak values of F_2 are generated in the central region of the composite couplers, but their values are null at both ends. The longitudinal shear stresses in adhesive τ_{xr}^a in **Figure 12** illustrate the antisymmetric characteristic along the bond length. It can be seen that the optimum fiber angle $\emptyset_2$ is 90°. This fiber orientation delivers the lowest maximum τ_{xr}^a/p_i of 0.6. **Figure 13** shows the radial normal stress in the adhesive. Interestingly, the values of σ_r^a are reduced by four to five times compared to the internal pressure applied over the whole range of the fiber angles considered.

The normal stresses σ_{11} and σ_{33} of adherend part 2 in the principal material coordinate system, when the fiber orientation are equal to 30°, are displayed in **Figure 14**. It can be noticed that under the application of the uniform internal pressure with 1 MPa magnitude, σ_{11} is

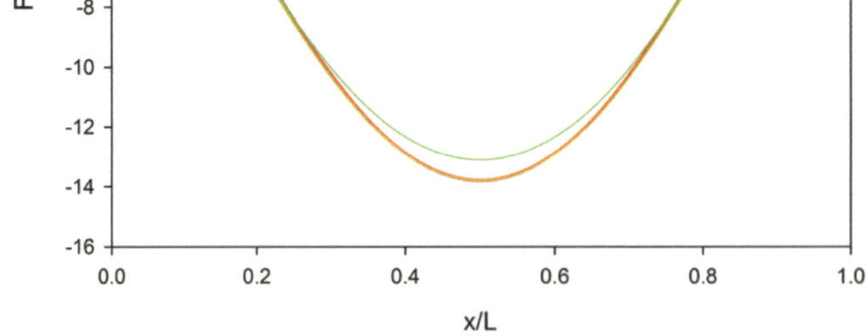

Figure 11. Force distribution in adherend part 2 of composite coupler joint subjected to internal pressure.

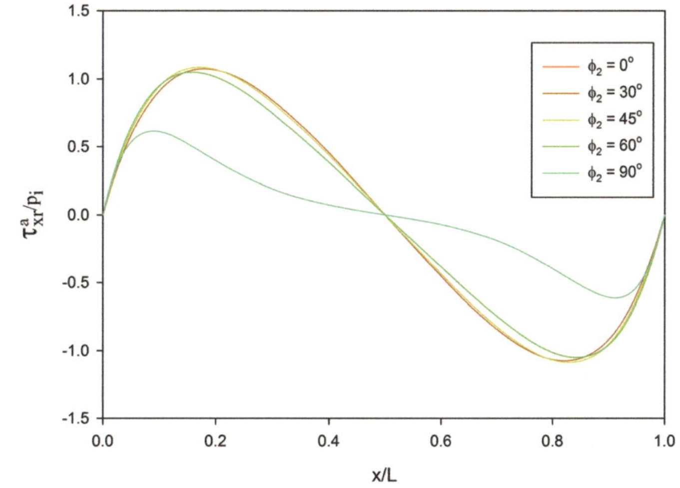

Figure 12. Adhesive axial shear stress distribution of composite coupler joint subjected to internal pressure.

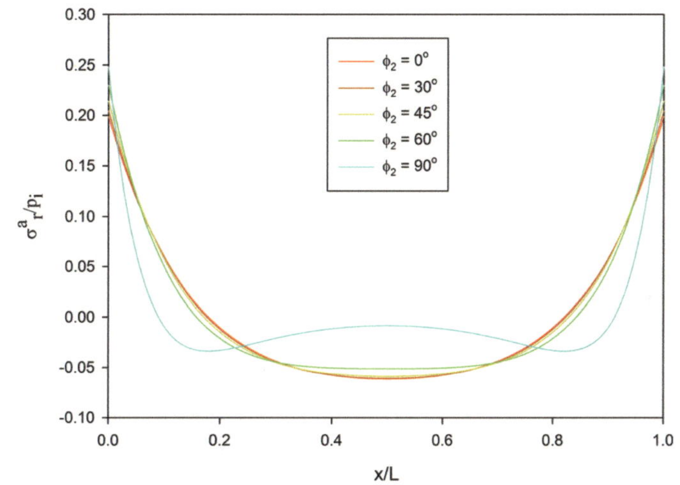

Figure 13. Interfacial radial stress distribution of composite coupler joint subjected to internal pressure.

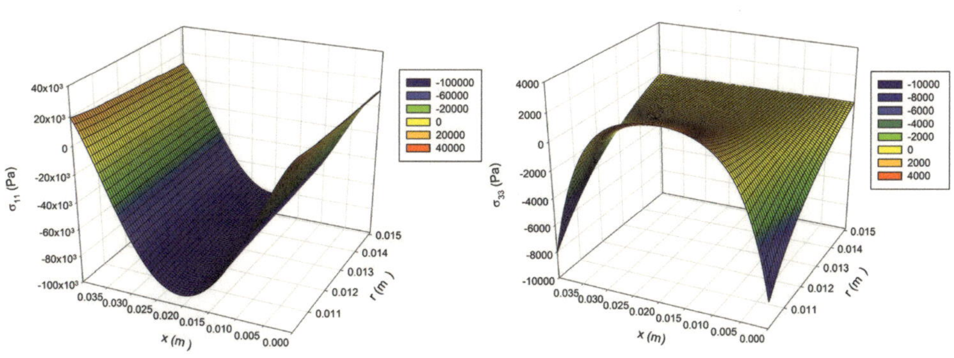

Figure 14. Distribution of stresses in composite coupler subjected to internal pressure.

maximum at the mid-length of bonding region, whereas σ_{33} is peak at $x = 0$ and 40 mm on the adhesive-adherend interface.

5. Conclusions

A unified mathematical model for predicting the joint stresses of the adhesive-bonded tubular-coupler joints or the equivalent bonded-lap joints under several types of load is formulated. The inner and outer adherends can be considered as an isotropic material, orthotropic material, or a laminated composite, whose fiber angle is constant along the tube axis. They are modeled as three-dimensional body and satisfied the equilibrium, kinematic, and constitutive equations in theory of elasticity. The adhesive is only treated to be a very thin isotropic elastic material with relative low modulus, and thus, merely three out-of-plane stress components are present. The finite-segment method is developed to compute adherend stresses in each small portion of the coupler. The analytical results obtained indicate the viability of the model for many joint conditions and configurations. The model can be used conveniently in the preliminary process of the joint design, which is usually critical in huge, complex, or integrated structures.

Author details

Sontipee Aimmanee

Address all correspondence to: sontipee.aim@kmutt.ac.th

Advanced Materials and Structures Laboratory (AMASS), Department of Mechanical Engineering, Faculty of Engineering, King Mongkut's University of Technology Thonburi (KMUTT), Bangmod, Bangkok, Thailand

References

[1] Lubkin JL, Reissner E. Stress distribution and design for adhesive lap joints between circular tubes. Transactions of ASME. 1956;**78**:1213-1221

[2] Adams R, Peppiatt N. Stress analysis of adhesive-bonded tubular lap joints. Journal of Adhesion. 1977;**9**:1-18

[3] Allman DJ. A theory for elastic stresses in adhesive bonded lap joints. The Quarterly Journal of Mechanics and Applied Mathematics. 1977;**30**:415-436

[4] Shi YP, Cheng S. Analysis of adhesive-bonded cylindrical lap joints subjected to axial load. Journal of Engineering Mechanics. 1993;**119**(3):584-602

[5] Nemes O, Lachaud F, Mojtabi A. Contribution to the study of cylindrical adhesive joining. International Journal of Adhesion and Adhesives. 2006;**26**:474-448

[6] Kumar S, Khan MA. An elastic solution for adhesive stresses in multi-material cylindrical joints. International Journal of Adhesion and Adhesives. 2016;**64**:142-152

[7] Kumar S. Analysis of tubular adhesive joints with a functionally modulus graded bondline subjected to axial loads. International Journal of Adhesion and Adhesives. 2009;**29**:785-795

[8] Volkersen O. Recheraches sur la theorie des assemblages colles. Construction Metallique. 1965;**4**:3-13

[9] Pugno N, Surace G. Tubular bonded joint under torsion: Theoretical analysis and optimization for uniform torsional strength. Journal of Strain Analysis. 2001;**36**(1):17-24

[10] Xu W, Li G. Finite difference three-dimensional solution of stresses in adhesively bonded composite tubular joint subjected to torsion. International Journal of Adhesion and Adhesives. 2010;**30**:191-199

[11] Nonlinear OJH. Analysis of adhesive bonded tubular single-lap joints for composites in torsion. Composites Science and Technology. 2007;**67**:1320-1329

[12] Spaggiari A, Dragoni E. Regularization of torsional stresses in tubular lap bonded joints by means of functionally graded adhesives. International Journal of Adhesion and Adhesives. 2014;**53**:23-28

[13] Aimmanee S, Hongpimolmas P. Stress analysis of adhesive-bonded tubular-coupler joints with optimum variable-stiffness composite adherend under torsion. Composite Structures. 2017;**164**:76-89

[14] Terekhova LP, Skoryi IA. Stresses in bonded joints of thin cylindrical shells. Strength of Materials. 1972;**4**(10):1271-1274

[15] Baishya N, Das RR, Panigrahi SK. Failure analysis of adhesively bonded tubular joints of laminated FRP composites subjected to combined internal pressure and torsional loading. Journal of Adhesion Science and Technology. 2017;**31**(19-20)

[16] Zhang Y, Qin TY, Noda, NA, Duan ML. Strength analysis of adhesive joints of riser in deep sea environment loadings. Applied Adhesion Science. 2013;**1**:1-9

[17] Apalak MK. Stress analysis of an adhesively bonded functionally graded tubular single lap joint subjected to an internal pressure. Science and Engineering of Composite Materials. 2006;**13**(3):183-211

[18] Herakovich CT. Mechanics of Fibrous Composites. USA: John Wiley& Sons, Inc; 1998

[19] Carlucci D, Payne N, Mehmedagic I. Small Strain Compatibility Conditions of an Elastic Solid in Cylindrical Coordinates. New Jersey: U.S. Army ARDEC; 2013. pp. 1-12

[20] Aimmanee S, Hongpimolmas P, Ruangjirakit K. Simplified analytical model for adhesive-bonded tubular joints with isotropic and composite adherends subjected to tension. International Journal of Adhesion and Adhesive (submitted)

Chapter 3

Development and Application of Low-Temperature Curable Isotropic Conductive Adhesive Toward to Fabrication in IoT Generation

Yusuke Okabe

Additional information is available at the end of the chapter

http://dx.doi.org/10.5772/intechopen.72662

Abstract

Flexible electronics is the expected technology in the future, and the bonding material may also require flexibility. Silyl terminated poly-ether (STPE) is a promising material that has both flexibility and low-temperature curability. In combination with tri-block polymer-based stretchable conductive paste and artificially formed fillet formed by elastic resin, it can build simple stretchable bonding system. It will be a prominent technology to satisfy the characteristics required by the near future devices.

Keywords: isotropic conductive adhesive, low-temperature bonding, flexible devices, stretchable devices, IoT

1. Introduction

Internet of Things (IoT) means all things are connected to the Internet showing the possibility of changing our lives. With the smartphone market becoming steady once, the explosive popularization of smartphones pushed for the miniaturization of electronic components and the spread of wireless Internet. In response to that, the wearable device approached a more practical device. Then, the role of electronics is becoming different from that of the previous one. For example, the reduction of medical expenses by health-care monitoring, efficient use of energy, assistance of workers and disabled people, and so on is said to be an important social task that electronics can solve. It is a great opportunity for printed electronics to join with the flow of giving an electrical function to various things indicated by IoT and create a light and soft device by printing. Attempts to design interfaces between the Internet world and real society have begun, as sensors and other electronic devices are incorporated all over our lives. In the trend,

IoT gives "anywhere anything" according to electric function indicated by IoT, base material has expanded from conventional rigid PCB to heat-sensitive materials, such as plastic, paper, and cloth, which can be formed by flexible circuits. In addition, those materials can be cured by low temperature, which is a major topic and important element technology. In fact, flexible circuits have been created by various materials and have achieved great results. To date, the softening of the wiring materials is largely developed by graphene in [1], carbon nanotubes in [2], silver nanowires in [3], conductive polymers in [4], or metal dispersed elastomers in [5–7]. With the advent of these technologies, flexible electronics can be said to have undergone a dramatic evolution. However, it has been rarely reported about the bonding materials, which coexist with flexibility and low-temperature curability. Furthermore, at present, it is difficult to create ICs, memories, communication modules, and so on by printing; therefore, it is necessary for the next generation devices to use hard parts and flexible substrates together. Additionally, in the connection between flexible substrates and hard components such as a wearable device, the connecting part will bear a large strain (**Figure 1**). As **Figure 2** shows, in Ref. [8], elongation that can occur on human body is around 30%. It is difficult to say that the bonding materials satisfy the requirement of imparting an electrical function to various substrates. In this situation, isotropic conductive adhesives (ICAs) have low-temperature curability and flexibility can be of great potential in creating next generation devices (**Figure 3**).

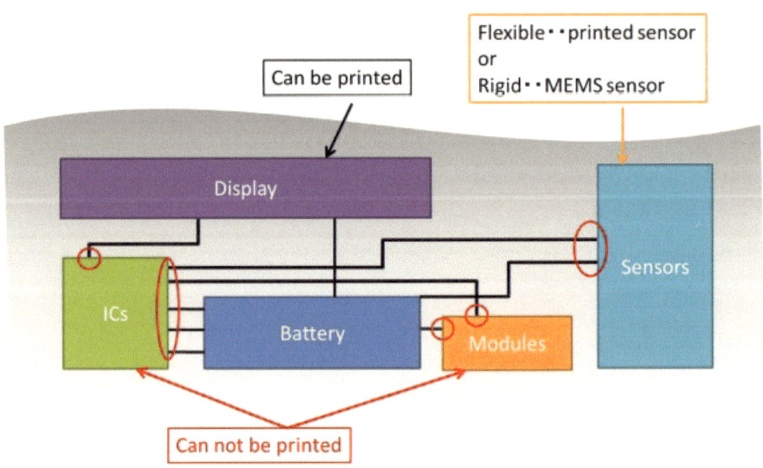

Figure 1. Image of flexible hybrid electronics device in future.

Parts	Elongation(%)
elbow(as flexion)	6-17
shoulder to elbow(as flexion)	13-34
shoulder(as rotation or stretch)	8-36
waist to hip(as flexion)	10-22
knee(as flexion)	2-51

Figure 2. Elongation of each part occurring in the human body.

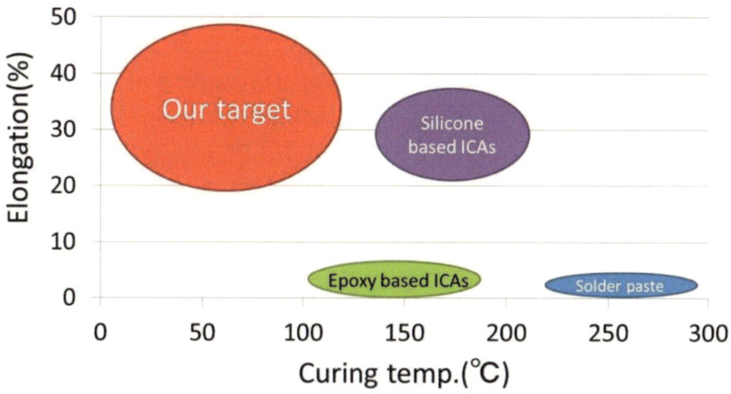

Figure 3. A comparison of this work with conventional conductive materials.

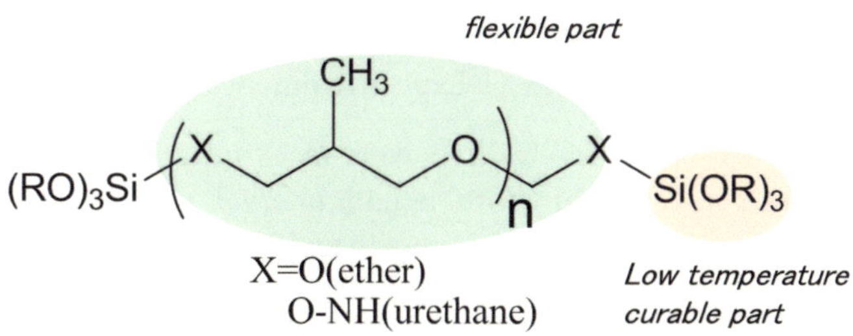

Figure 4. Chemical structure of STPE.

As the types of base materials are expanding, the most important basic property is to ensure adhesion to various base materials, and the adhesion technology is meaningful.

Silyl terminated poly-ether (STPE), as shown in **Figure 4**, is flexible and strong material to repeated strain, such as heat cycle. In present, there are many achievements to sealant for construction and adhesive for electronic components. Conventional conductive adhesives are mainly thermosetting system; the heat energy required for curing itself has disturbed the applications of heat-sensitive substrates and components. The cross-linkable silyl group can react by moisture in the air, even at room temperature, and it is suitable to implement thermally sensitive components. That is, it is possible to assemble the electronic component on the film as well as paper, fabrics, or various materials. We have developed a conductive adhesive having low-temperature curableness and flexibility by distributing the silver microsized to STPE. Here, we describe the characteristics of new isotropic conductive adhesive based on STPE, which has low-temperature curableness and flexibility and its application.

2. Characteristics of the STPE based conductive adhesive

First, we describe the design and characteristics of new conductive adhesives. Next, we describe the low-temperature curability, flexibility, and stretchable conductive paste for flexible devices, and finally, we introduce their application.

2.1. Experimental section

ICAs were fabricated by uniformly dispersing microsized silvers as described later. Uniform dispersion was achieved by a high-speed blender operated at 2000 rpm for 3 min under vacuum conditions (ARV-310, THINKY Company). Epoxy-based ICAs were formulated by Epicote 828 (Mitsubishi Chemical Co., LTD.) and 2-ethyl-4-imidazole (Tokyo Chemical Industry Co., LTD.) as curing agents. ICAs were mask-printed onto glass plate. STPE-based ICAs were cured at 80°C for 2 h, and epoxy-based ICAs were cured at 120°C for 1 h. Their dimensions were 80, 100, and 0.5 mm in width, length, and thickness, respectively. Volume resistivity was measured by MCP-T360 (Mitsubishi Analytec Co., LTD.).

Adhesion strength is tested by die shear tester (Dage 4000) at a shear head speed of 500 μm/s. Epoxy-based ICA (XA-874, FjikuraKasei Co., LTD.) was used as comparison. ICAs were mask-printed onto copper plate on printed circuit board and mounted on 3216-sized chip resistor. Printed thickness was 0.1 mm. The high temperature and high humidity tests were carried out at 85°C and 85% relative humidity (RH), respectively. Heat cycle test was performed at the temperature range between −40 and 105°C. Exposure time was 30 min.

Electrical stability is tested on ISO-16525. ICAs were mask-printed onto copper electrodes, in which dimensions were 2, 4, 8 mm in width, length, interval, respectively, on printed circuit board. Dimensions of ICAs were 4, 100, and 0.1 mm in width, length, and thickness, respectively.

Stretchable conductive paste was fabricated by uniformly dispersing microsized silver flakes. Uniform dispersion was achieved by a high-speed blender operated at 2000 rpm for 3 min under vacuum conditions (ARV-310, THINKY Company). Conductive pastes were mask-printed onto various substrates and dried at 100°C for 30 min. Their dimensions were 80, 100, and 0.5 mm in width, length, and thickness, respectively. Cross cut test was performed based on ISO-2649.

Bending resistance test of conductive paste was performed on IEC62715 by DLDMLH-4U (YUASA SYSTEM Co., LTD.). Conductive pastes were mask-printed onto PET film (Lumirror S10, TORAY Co., LTD.) and dried at 100°C for 30 min. Their dimensions were 5, 100, and 0.2 mm in width, length, and thickness, respectively. To investigate the resistance change in real time, copper and lead were connected to both ends, and the resistance was measured in current of 100 mA by the four-terminal method.

Dynamic boding resistance test was similar to the abovementioned test. Copper foiled polyimide was used as circuit. XA-874 and solder paste (FLF01-BZ(L), Matsuo HANDA Co., LTD.) were used as comparison. Artificially formed fillet was formed by SuperXG No.777 (CEMEDINE Co., LTD.).

In stretched resistance test, conductive pastes were mask-printed onto TPU (Platilon VPT9122, Covestro Japan Co., LTD.) and dried at 100°C for 30 min. Their dimensions were 5, 100, and 0.2 mm in width, length, and thickness, respectively.

2.2. Design of STPE-based conductive adhesive

Figure 5 shows formulation and volume resistivity of conductive adhesive. In case of using STPE as base binder, it is understood that combining silver fillers having different shapes and TAP density is important in expressing conductivity. Also, urethane bond in the polymer backbone showed less resistivity. Hard segment derived from a hydrogen bond due to a urethane bond plays strengthening of the interaction and the conductive path between the filler. Otherwise, BPA-based conductive adhesive has higher conductivity when using a single shape. Previously, there are few cases mentioning the TAP density in the formulation of the conductive adhesive, and it has been discussed in the shape of fillers. However, these results show the selectivity of characteristic silver fillers to develop the conductivity to flexible binder as STPE (**Table 1**).

2.3. Curing behavior

Curing behavior of STPE-based conductive adhesive is shown in **Figure 6**. Curing proceeds even at room temperature, the conductivity increased with time. Further, by heating to

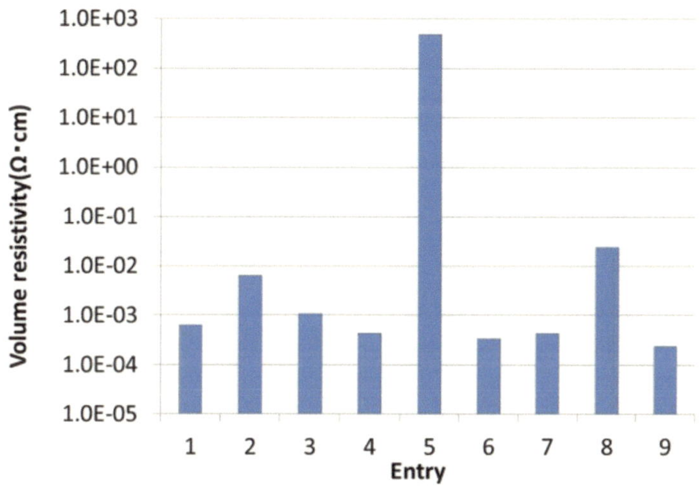

Figure 5. Volume resistivity of designed ICAs shown in **Table 1**.

Entry	TAP density of filler (g/cm^3)	3.5	1.0	1.5
	Base resin	**Silver flake**	**Aggregated silver**	**Spherical silver**
1	STPE (ether)	300	200	–
2	STPE (ether)	300	–	200
3	STPE (ether)	300	100	100
4	STPE (ether)	200	300	–
5	STPE (ether)	500	–	–
6	STPE (urethane)	300	200	–
7	BPA	500	–	–
8	BPA	300	200	–
9	BPA	–	500	–

Table 1. Formulation of ICAs, silver weight ratio (per 100 resin).

50 or 80°C, curing is accelerated and conductivity increases quickly. Conventional epoxy-based adhesives are often unable to exert their performance unless they adhere to the recommended curing conditions. In other words, the curing conditions themselves may cause trouble in bonding process. With STPE-based conductive adhesive, even if heating is stopped halfway, the reaction proceeds, so that, a flexible production process can be constructed. Also, die shear strength increased with time, curing is accelerated and die shear strength increased quickly. On the other hand, the expression of adhesion strength does not match with the conductivity; this is because the adhesion is rate-limiting between electrode and adhesive interface.

2.4. Adhesion durability

Figure 7 shows adhesion durability at high temperature and high humidity, and heat cycle (−40–105°C) compared with thermosetting epoxy resin-based electrical conductive adhesive that is conventionally used. Initial adhesion strength of the STPE-based ICA is smaller than the epoxy resin system, since STPE has a lower elastic modulus than conventional epoxy resin. For this reason, it is better to consider a device having a slightly different design philosophy such as a flexible device than a simple replacement for solder. In particular, STPE-based ICA exhibits excellent bonding strength retention under heat cycle environment. On the other

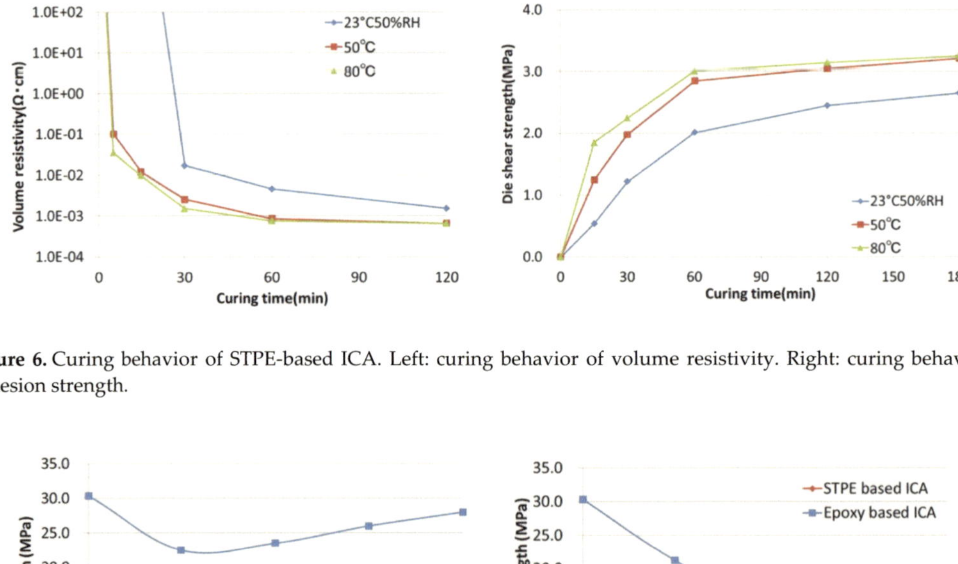

Figure 6. Curing behavior of STPE-based ICA. Left: curing behavior of volume resistivity. Right: curing behavior of adhesion strength.

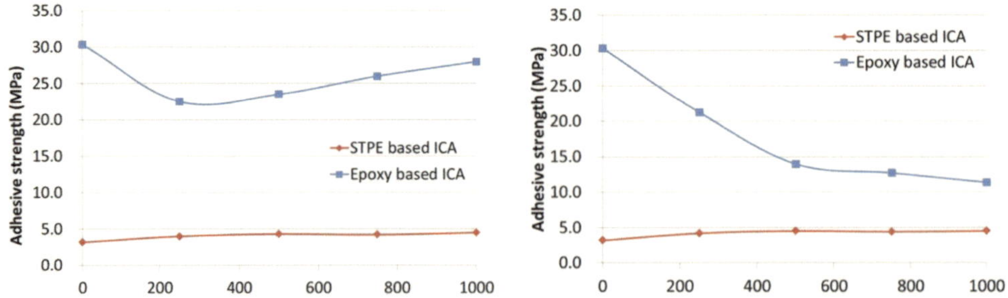

Figure 7. Adhesion strength at durability test. Left: 85°C and 85% RH. Right: heat cycle (−40–105°C).

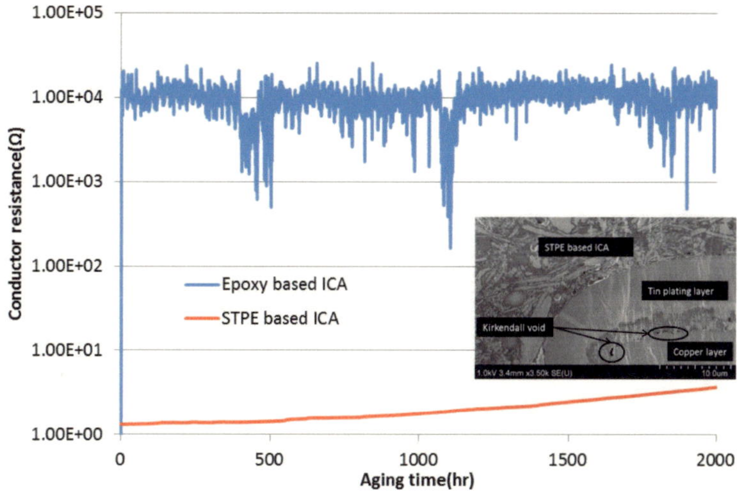

Figure 8. Conductor resistance of STPE-based ICA (red) and epoxy-based ICA (blue) at 85°Cand 85% RH on tin plated-copper electrode. SEM images show sectional view of PCB in the test of STPE-based ICA after 2000 h.

hand, the adhesion strength of epoxy-based ICA has decreased in every cycle. This phenomenon is induced by internal stress generated at the bonding interface when the temperature change occurs across the glass transition point in Refs. [9–11]. In this situation, rigid base resin cannot reduce the internal stress at bonding interface. STPE has low modulus and flexibility that induce stress relaxation characteristics, and it leads to ensure long-term reliability.

2.5. Electrical stability

The excellent characteristics of STPE-based ICA are shown in **Figure 8**. Galvanic corrosion between tin electrode and silver fillers in a high-temperature and high-humidity environment has become a long-standing problem of the epoxy-based ICAs. Therefore, in order to put the conductive adhesive into practical use, an increase in parts' cost has become a problem due to the use of a gold electrode or the like. Corrosion of the tin electrodes is said to be accelerated by the chloride ion in epoxy resin in Ref. [12]. In contrast, STPE is not containing chloride ion, and STPE-based ICA does not occur corrosion on tin electrode. A slight increase is observed in the conductor resistance after 1000 h, and it is found that it was generated by Kirkendall void between tin plating layer and copper used as electrode from SEM observation.

3. Characteristics of conductive paste

3.1. Design of flexible/stretchable conductive paste

As stated earlier, securing adhesion to enlargement of applied base material is an important factor, and adhesion technology is a key point. Conventional conductive paste does not have

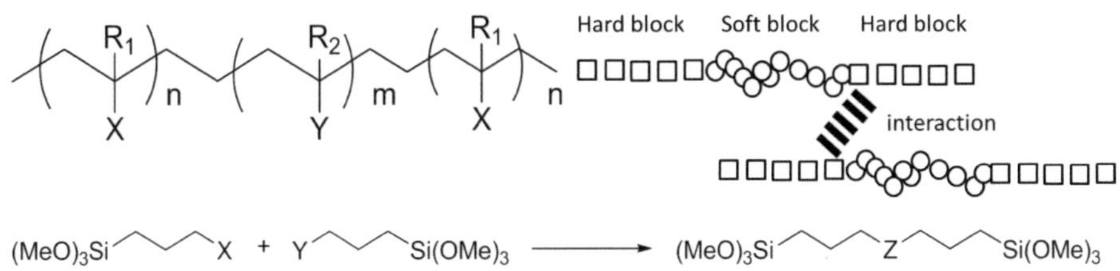

Figure 9. Schematic image of tri-block elastomer and pre-cured silane coupling agents.

Base polymer	Acrylic	SIS	SEBS
Volume resistivity	3.50E − 04	8.50E − 05	7.50E − 05
Adhesion to PET	100/100	70/100	75/100
Adhesion to PEN	100/100	60/100	70/100
Adhesion to TPU	100/100	50/100	70/100
Adhesion to excimer laser treated SR	100/100	0/100	0/100

Table 2. Test results of block polymer-based conductive paste.

extensive adhesion property. Therefore, we have to choose conductive paste according to substrates. To solve it, we choose tri-block elastomer as binder and pre-cured silane coupling agents (**Figure 9**).

Adhesion test results by cross cut test is shown in **Table 2**. Hydrocarbon polymers such as SIS and SEBS show high conductivity, but they do not exhibit extensive adhesion property. On the other hand, it was found that the acrylic polymer has good adhesion to various substrates and is excellent in balance with conductivity. In subsequent experiments, the conductive paste based on an acrylic polymer is used.

3.2. Dynamic durability of conductive paste

Figure 10 shows real time resistance change due to bending on PET film. The resistance change due to bending is smaller than that of conventional epoxy-based flexible conductive paste. In addition, the resistance fluctuation when bending once is also small, which is considered to be based on polymer with hard segment and soft segment coexisting.

Figure 11 shows the resistance change at elongation on TPU. For example, the strain accruing on the human body is about 50% at the maximum. It is thought that the hysteresis of resistance is small, and it functions as strain sensor, which is capable of detecting elongation of about 50%. Also, by utilizing high adhesion and flexibility, it is possible to form circuits for various substrates.

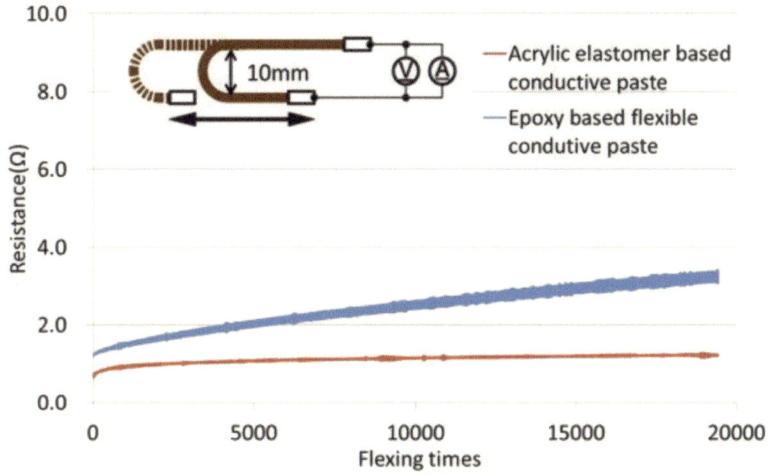

Figure 10. Conductor resistance change of acrylic elastomer-based conductive paste in bending test.

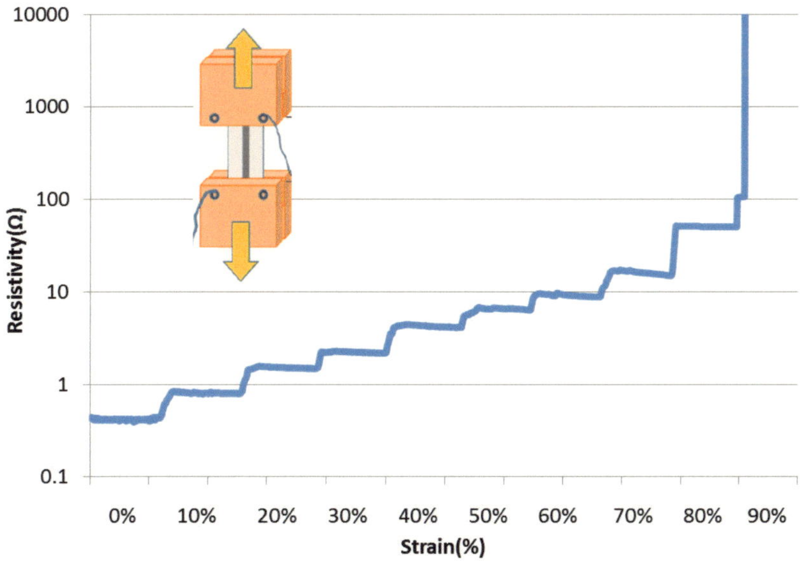

Figure 11. Conductor resistance change of acrylic elastomer-based conductive paste at elongation. Circuit size width: 5 mm, length: 100 mm, and thickness: 30 μm.

4. Application for flexible/stretchable bonding system

As mentioned earlier, STPE-based ICA does not have strong bonding strength as conventional materials. Therefore, in order to take advantage of this material, it is necessary to construct reinforcing structure. Solder or thermosetting conductive adhesive forms a fillet at the time of curing, and it becomes a reinforcing layer against dynamic strain. On the other hand, since STPE-based ICA is a low-temperature curable, it is difficult to form fillets. **Figure 12** shows bending resistance of mounting on FPC by STPE-based ICA and conventional materials and the effect of artificially formed fillet. Although the bonding resistance of solder is stable, bonding resistance of STPE-based ICA does not form a fillet, which has a considerable resistance variation. Even in the case of a thermosetting epoxy-based ICA, a slight resistance

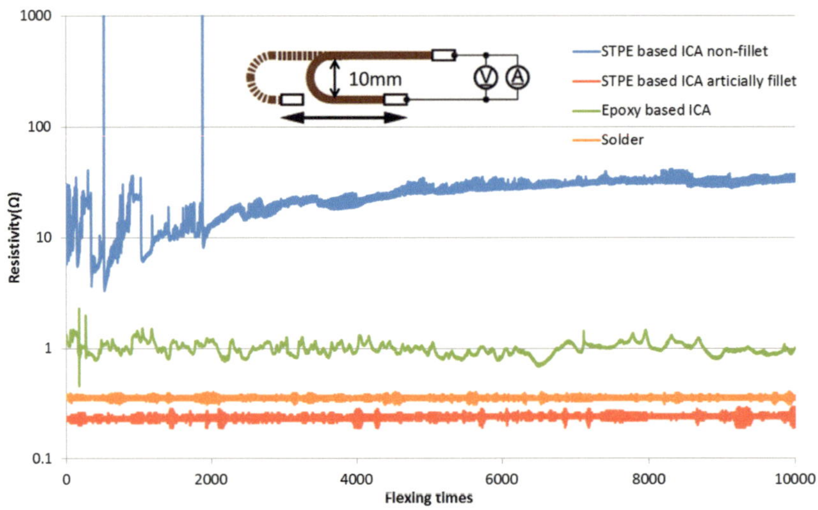

Figure 12. Bonding resistance change of various materials on FPC in bending test.

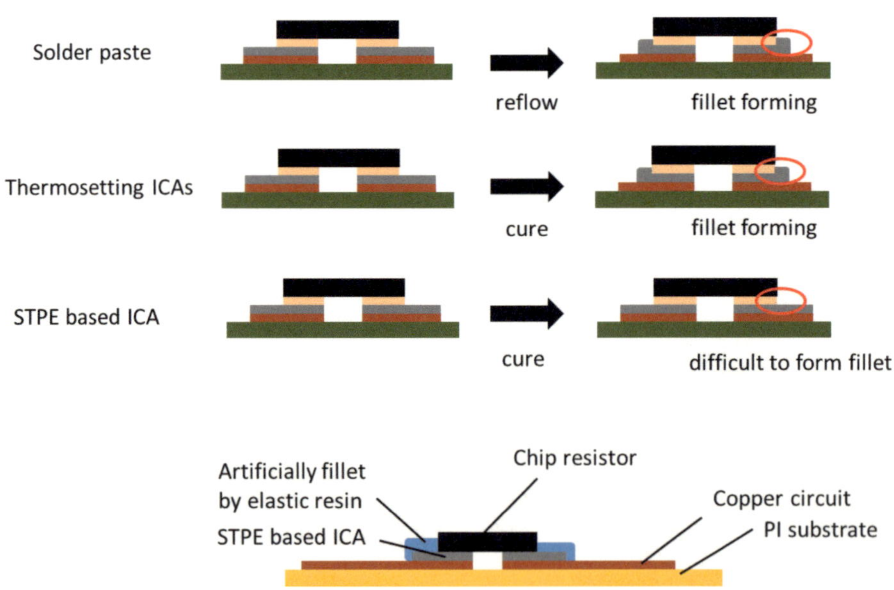

Figure 13. Images of fillet forming at bonding process and detailed image of reinforced STPE bonding structure.

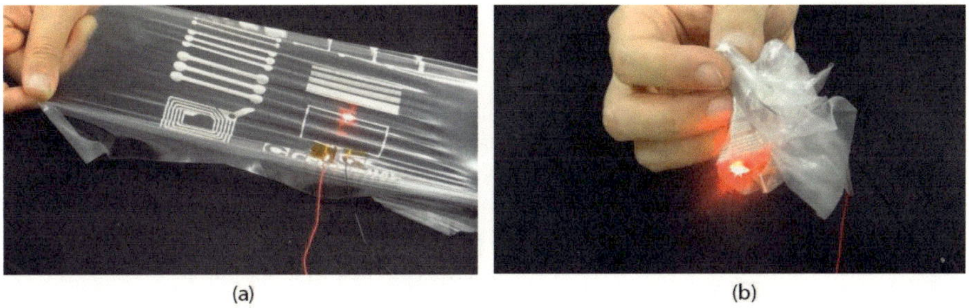

Figure 14. Pictures of implementation of LED on stretchable substrate. Left: stretched. Right: crumpled.

change is observed. On the other hand, fillet formed STPE-based ICA get bending resistance similar to solder connection.

Detailed structure is shown in **Figure 13**. This structure can also be applied to elastic substrates. Previously, there have been various reports on mounting of rigid parts on stretchable base materials, but it is mainly focused on design of base substrates and how to use rigid bonding materials [13, 14]. By combining STPE-based ICA and flexible reinforcing material, it is not necessary to design a complicated base material and it is possible to simply mount on a flexible base material such as **Figure 14**. In **Figure 14**, a wiring is drawn on a stretchable substrate with a stretchable conductor, and LED chip is mounted with STPE-based ICA and reinforced by elastic resin. Even without constructing a complicated mounting structure, disconnection is not observed, and LED chip continued to light up during repetition of expansion.

5. Conclusion

Characteristics of the conductive adhesive having both low-temperature curability and flexibility, which is based on STPE shows a great advantage compared with conventional conductive adhesive. In addition, it showed dynamic durability rivaled to conventional solder connection. Namely, the bonding system which is almost constructed by elastic resin and highly reliable is a promising technology for electronics in the next generation. Fabrication in IoT and wearable devices still has room for improvement. STPE-based ICA is developed as one solution for the future electronics society and we expect that innovation will happen.

Author details

Yusuke Okabe

Address all correspondence to: yusuke.okabe@cemedine.co.jp

Research and Development Division, CEMEDINE Co., LTD., Japan

References

[1] Qin Y, Peng Q, Ding Y, Lin Z, Wang C, Li Y, et al. Light weight, superelastic, and mechanically flexible graphen/polyimide nano composite foam for strain sensor application. ACS Nano. 2015;**9**:8933-8941. DOI: 10.1021/acsnano.5b02781

[2] Yu Y, Luo S, Sun L, Wu Y, Jiang K, Li Q, et al. Ultra stretchable conductors based on buckled super-aligned carbon nano tube films. Nanoscale. 2015;**7**:10178-10185. DOI: 10.1039/C5NR01383F

[3] Kim KK, Hong S, Cho HM, Lee J, Suh YD, Ham J, et al. Highly sensitive and stretchable multidimensional strain sensor with prestraind anisotropic metal nanowire percolation network. Nano Letters. 2015;**15**:5240-5247. DOI: 10.1021/acs.nanolett.5b01505

[4] Bandodkar AJ, Nuñez-Flores R, Jia W, Wang J. All-printed stretchable electrochemical devices. Advanced Materials. 2015;**27**:3060-3065. DOI: 10.1002/adma.201500768

[5] Araki T, Nogi M, Suganuma K, Kogure M, Kirihara O. Printable and stretchable conductive wiring comprising silver flakes and elastomer. IEEE Electron Device Letters. 2011;**32**:1424-1426. DOI: 10.1109/LED.2011.2161663

[6] Sekitani T, Noguchi Y, Hata K, Fukushima T, Aida T, Someya T. A rubberlike stretchable active matrix using elastic conductors. Science. 2008;**321**:1468. DOI: 10.1126/science.1160309

[7] Matsuhisa N, Kaltenbrunner M, Yokota T, Jinno H, Kuribara K, Sekitani T, et al. Printable elastic conductors with a high conductivity for electronic textile applications. Nature Communications. 2015;**6**:7461. DOI: 10.1038/ncomms8461

[8] Aratani Y, Kojima T. Stretch fabrics. Sen-ito Kogyo. 1984;**40**(4-5):318-321. DOI: 10.2115/fiber.40.4-5_P318

[9] Devroy DRE, Homma M. Blends of silyl-terminated polyethers and epoxides as elastic adhesives. International Journal of Adhesion and Adhesives. 2001;**21**(4):275-280. DOI: 10.1016/S0143-7496(00)0035-X

[10] Berzins R, Meri RM, Zicans J. Compatibilizers effect on silyl-terminated polyether/epoxy resin system mechanical ans rheological properties. Key Engineering Materials. 2017;**721**:441-445. DOI: 10.4028/www.scientific.net/KEM.721.441

[11] Hashimoto K, Imaya K. Silyl-terminated polyethers for sealant use: Performance updates. Adhesives Age. 1998;**41**(8):18-22. ISSN: 0001-821X

[12] Kim S-S, Kim K-S, Lee KJ, Kim S, Suganuma K, Tanaka H. Electrical resistance and microstructural changes of silver-epoxy isotropic conductive adhesive joints under high humidity and heat. Journal of Electronic Materials. 2011;**40**(2):232-238. DOI: 10.1007/s11664-010-1421-z

[13] Kubo M, Li X, Kim C, Hashimoto M, Wiley BJ, Ham D, Whitesides GM. Stretchable microfluidic radiofrequency antennas. Advanced Materials. 2010;**22**:2749-2752. DOI: 10.1002/adma.200904201

[14] Hu X, Krull P, de Graff B, Dowling K, Rogers JA, Arora WJ. Stretchable inorganic-semiconductor electronic systems. Advanced Materials. 2011;**23**:2933-2936. DOI: 10.1002/adma.201100144

Section 2

Wood Adhesive Bonding

Chapter 4

Green Binders for Wood Adhesives

Emelie Norström, Deniz Demircan,
Linda Fogelström, Farideh Khabbaz and
Eva Malmström

Additional information is available at the end of the chapter

http://dx.doi.org/10.5772/intechopen.72072

Abstract

Today's society relies heavily on glued wood products for constructions, furniture, and floorings, for example. Essentially, all adhesives on the market are based on fossil-based resources, and many also contain formaldehyde to yield sufficient reactivity and adhesive performance. Formaldehyde is soon to be banned from consumer goods in Europe, due to its carcinogenic and allergenic features. With the rapidly growing societal environmental awareness, it becomes evident that it is crucial to seek greener, more sustainable alternatives. There is nothing new to this idea; on the contrary, prior to the advent of synthetic polymers, a range of biopolymers such as proteins and starch, were successfully used. However, since adhesives based on synthetic polymers were found to perform better, especially regarding the water resistance, the naturally sourced adhesives have had a subordinate role up until recently. The growing interest for using bio-polymers from renewable resources, such as wood/forest, corn, and cereals have spurred significant R&D developments toward the use of bio-polymers in green wood adhesives. The scope of the present chapter is to summarize, in short, some of the most recent scientific literature regarding the development of green adhesives.

Keywords: wood adhesive, protein, polysaccharide, starch, chitosan, hemicellulose, tannin, lignin

1. Introduction

Wood adhesives are produced in large amounts and are used for many large-scale applications such as load-bearing constructions, flooring, furniture, doors, and windows. Today, wood adhesives are essentially solely prepared from fossil-derived polymers based on, for example, urea, melamine, formaldehyde, phenol, resorcinol, isocyanate, and vinyl acetate, but historically adhesives were prepared from various natural sources, such as proteins from milk, blood, and soybean [1]. In 1928, Casco Adhesives (now AkzoNobel Adhesives) in Sweden, started their production of adhesives based on casein (Casein Company), a milk protein [2]. With the rapid and revolutionizing development of the "plastic era," initialized during the 1940s with the launch of synthetic polymers such as acrylates and vinyl esters, it became apparent that the performance of bio-based adhesives could be widely exceeded. During the 1960s, the hitherto used natural binders were replaced by synthetic polymers derived from fossil-based resources, motivated both by insufficient properties and high cost [3, 4].

Fossil-derived adhesives are cost-effective and perform very well regarding bonding performance and water resistance. The oil-crisis during the 1970s brought about a realization that natural sources for fossil-based products are limited; however, this has not yet had a dramatic effect on the adhesive industry. A large share of the industrially viable wood adhesive systems also comprises formaldehyde which is a highly reactive compound, making it well suited for its intended use. For instance, particleboard products are almost exclusively bonded with formaldehyde, in combination with urea and/or melamine or phenol. However, formaldehyde has been identified as a very hazardous compound and will most likely be banned from use in many applications [5]. The formaldehyde ban, combined with the rapidly growing environmental awareness calls for the adhesive industry to become more sustainable, more benign, and less fossil-dependent. To replace fossil-based adhesives with sustainable counterparts successfully, the adhesive properties and bonding performance have to be very similar, or bring other added values, and very importantly, the cost performance has to be on par with existing, non-sustainable, adhesives.

The urge for developing green adhesives reawakens the interest for bio-based adhesives, even though it is not completely uncomplicated. Many natural resources such as starch and protein, potentially well-suited for adhesive applications, may also be used as food sources. This calls for careful consideration in a time where more than 800 million people suffer from starvation world-wide [6].

The development of bio-based adhesives also poses other challenges; adhesives prepared from bio-based polymers often exhibit poor water resistance and/or render too high cost to successfully compete with fossil-derived polymers [1]. Another challenge when utilizing bio-based polymers is the often large property variation, emanating from locus and constantly shifting growth-conditions such as type of source, growth season, access to nutrients, climate, etc. [7]. Also, the extraction and fractionation procedures required to isolate the bio-based polymers will influence the final properties of the adhesive, as well as the cost. Altogether, it is challenging to replace robust synthetic polymers, having well-known and reproducible characteristics, with bio-based polymers with a broader property window, in industrial applications. Today, the research in this area is mainly focused on proteins, starch and other polysaccharides, lignin, and tannin as raw materials (**Figure 1**).

1.1. Evaluation of bonded specimens

Wood adhesives can be applied on several different substrates such as veneers, plywoods, panels, beams, particleboards, etc., and are often subjected to elevated temperatures and high

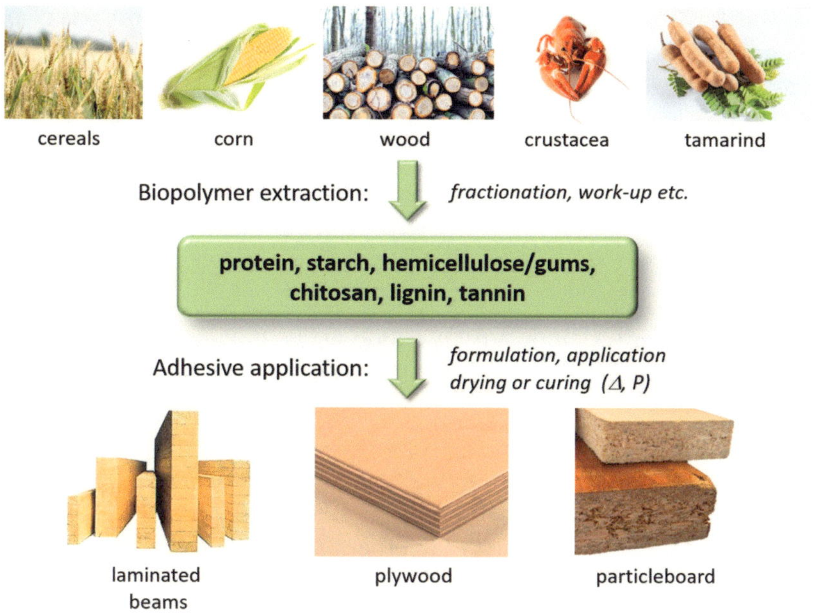

Figure 1. Green binders for wood adhesives. Photocredits for pictures of particleboard and plywood distributed by a CC-BY-SA 3.0 license [8].

pressure during cure depending on the application. Adhesives are evaluated by testing the bond strength of the bonded wood specimens; both dry and wet strength are crucial properties. The requirement of the adhesive depends on the application (e.g., interior or exterior use), type of adhesive (e.g., thermoplastic or thermosets), type of wood substrate, etc. Heat resistance is of particular interest for specimens glued with thermoplastic adhesives, since a thermoplastic material generally is more sensitive to heat than the thermosetting counterpart. There are many different national standardized evaluation procedures, such as ASTM, EN, Chinese industry standard, China National Standards, etc., with different criteria for fulfilling the requirements of a certain application. These standards often differ in the gluing procedure, sample preparation, pressing conditions, conditioning and performance evaluation. The use of slightly dissimilar standards and evaluation protocols make it difficult to compare different studies [9, 10]. The evaluation of mechanical properties of wood adhesives has been subject to substanitial research [11].

The chemical composition of adhesives is often characterized with Fourier transform infrared spectroscopy (FTIR), nuclear magnetic resonance (NMR), and the relative average molecular weight is determined by size exclusion chromatography (SEC). Further characterization includes rheological studies (mainly the viscosity), thermal studies using differential scanning calorimetry (DSC) and thermogravimetric analysis (TGA), contact angle measurements, and storage stability measurements. The bond line and the adhesive's penetration into the wood in the bonded specimens can be studied with optical microscopy (OM), scanning electron microscopy (SEM), transmission electron microscopy (TEM), X-ray photoelectron spectroscopy (XPS), and energy dispersive X-ray spectroscopy (EDXS), etc.

The scope of this book chapter is to give a short overview of the most recent publications concerning the use of bio-polymers as promising sustainable resources for wood adhesives. Several extensive and in-depth reviews of the use of green wood adhesives are already published and these are referred to throughout the chapter, when appropriate. The chapter does not cover patents or scientific publications published in any other language than English.

2. Protein

Proteins are linear polyamides built up by amino acids, linked together with polypeptide bonds, and are, together with DNA, fat, and polysaccharides, the most important constituent in all living species. There are 20 different amino acids, either acidic, basic, or neutral depending on the structure of the side chain. The properties of a protein originates from its complex structure; the amino acid sequence (primary structure) is partly arranging into α-helices and β-sheets (secondary structure), side-chains interact to form a 3D-structure (tertiary structure), and the whole protein molecule interacts with other protein molecules to form a higher-order (quaternary) structure [12].

Proteins have a long history as binders in wood adhesives but were replaced by fossil-based polymers due to cost and insufficient properties, such as poor bond strength and water resistance [1]. The side-chains of the polypeptide chain contain functional groups that make the amino acid either hydrophilic or hydrophobic, and provide possibilities for interaction with hydroxyl or carboxyl groups in wood, and for crosslinking.

Generally, protein adhesives suffer from high viscosity, consequently demanding low solid content, and they commonly only meet the requirements for indoor applications due to poor water resistance [13]. Extensive research is being conducted to improve the bonding performance and water resistance of proteins to extend the applicability of wood-bonded protein adhesives, as previously reviewed in literature [13–15].

Soybean protein can be obtained from soybeans during the production of soy oil and soy meal. Soy oil is used in the food industry and soy meal as animal feed [16]. Soy protein is a promising alternative to fossil-based adhesives due to its availability, easy processing, and low cost, but the fact that it is a food source is debatable. Soybean protein-based adhesives have been suggested in a few applications today: particleboards, laminated plywood, and finger-joint lumber [13, 17]. So far, the use of soybean protein in adhesives has been limited due to low water resistance and high viscosity [13]. Adhesives containing soybean flour is commercially available [18].

Physical and chemical methods have been used to improve the properties of proteins. Denaturation of the native protein structure exposes functional groups buried within the 3D-structure of the protein, which may enable solubilization and bonding. The increased solubilization further makes it possible for the protein adhesive to flow better over the wood surface, forming hydrogen bonds with wood, and allows for subsequent chemical crosslinking. Denaturisation can be triggered by increased temperature, pressure, and changes in pH, as well as the addition of denaturants, such as urea guanidine hydrochloride, enzymes, SDS or other detergents [13, 19].

It was shown that a combination of thermal alkali degradation, thermal acid treatment, and crosslinking of soybean protein can provide an adhesive with better applicability regarding viscosity and bonding performance. Thermal alkali degradation lowers the viscosity and improves the technological applicability. Thermal acid treatment allows for the formation of an intermolecular network which improves the water resistance. Glyoxal, epoxy latex, polyisocyanate, and polyamide were evaluated as crosslinkers. The type of crosslinker, amount of crosslinker, and ratio of thermal alkali-degraded soybean protein to thermal acid-treated soybean protein, had important effects on the final adhesive properties. Polyamide is the preferred crosslinker due to its good crosslinking efficiency, miscibility with soybean protein, and low viscosity [20].

Thermo-chemical treatment of soybean protein in the presence of sodium sulfide or sodium dodecyl sulfate, followed by crosslinking with epichlorohydrin-modified polyamide (EMPA) showed promising results. The adhesive could withstand recurring hygrothermal treatment after which the wet strength fulfilled the required value for structural use [21].

Proteins have also been combined with synthetic polymers and resins such as formaldehyde, urea formaldehyde (UF), melamine urea formaldehyde (MUF), and dispersion polymers, such as polyvinyl acetate (PVAc) [14, 22]. The water resistance of protein adhesives can be improved by the addition of crosslinkers such as polyamidoamine-epichlorohydrin (PAE) resin or polymeric diphenylmethane diisocynate (pMDI) [14].

In a recent study, water was substituted with egg white in a soybean-meal adhesive which was subsequently crosslinked with triglycidyl amine in an attempt to increase the solid content and to improve the water resistance of soybean protein-based adhesives. As a result, the solid content was increased, the viscosity was kept low, and the bonding performance was improved. The wet strength of the adhesive was superior to that of conventional UF resin and PAE-crosslinked soybean protein-based adhesive, and was comparable to that of melamine-modified UF resin [23].

A green route to prepare a soybean protein-based adhesive has been demonstrated by utilizing the polysaccharides and proteins in soy meal. First, the polysaccharides and proteins were separated, whereafter the polysaccharides were crosslinked with a green crosslinker, sodium hexametaphosphate, and subsequently blended with the proteins to form an interpenetrating network. After hot pressing, a stable glue line was formed that fulfills the requirements for plywood in interior use [24]. In another study, the polysaccharides were hydrolyzed and then crosslinked with the proteins through a Maillard reaction. The bonding performance, rheological properties, and thermal stability were improved and the adhesive also met the requirements for plywood in indoor applications [25].

Magnesium oxide (MgO) has been added to soybean protein-based adhesives to improve the bonding performance. In a recent study, MgO was used together with different fractions from soybean. It is preferable to use soy flour or soy meal directly without purification steps which would increase the cost of the final product. However, the purified soybean protein gave better adhesive properties together with MgO, compared with the polysaccharide-containing soy flour [26].

The water resistance of a soybean protein-based adhesive can be improved with the addition of 5,5-dimethyl hydantoin polyepoxide (DMHP). DMHP acts as a crosslinker during cure while at the same time decreasing the wet-adhesive viscosity, improving the wetting and penetration of the adhesive, yielding a smooth surface and bonding performance. DMHP made it possible to increase the solid content, decrease the viscosity, and improve both dry and wet strength [27].

A new type of soybean protein-based adhesive has been presented where a soybean protein-acrylate emulsion-based adhesive was synthesized by mini-emulsion polymerization. Methyl methacrylate and butyl acrylate were polymerized using soybean protein as the protective colloid, resulting in an adhesive evaluated for plywood preparation. The bonding performance was, however, only slightly improved [28].

Soybean protein has been blended with ethylene glycol, diethylene glycol, and polyethylene glycols with different molecular weights. Ethylene glycol improved the wet strength of the adhesive which was suggested to be due to improved wettability and hydrogen bonding. Higher molecular weight polyols, however, decreased the wet strength of the adhesives [29].

Soybean protein has also been blended with lignin to improve the water resistance of the adhesive [30]. The lignin-particle size and the protein-to-lignin ratio significantly affected the bonding performance. Lignin with smaller particle size increased the wet strength of the adhesive [31]. Lignin amine was prepared by Fenton oxidation and subsequent reductive amination [32]. Addition of lignin amine to soybean protein-based adhesives has also shown to improve the bonding performance and water resistance.

Soybean protein and cottonseed protein isolates were mixed in different ratios and used as wood adhesives. Increasing fraction of soybean protein was deteriorating the adhesive properties. However, formulations with addition of ca 50% xylan, starch, or cellulose exhibited a retained hot-water resistance [33].

Soy crops are unfortunately not widespread over the world; therefore, the possibility of using other types of proteins has also been explored. Wheat gluten is a by-product from the production of wheat starch and bioethanol, and has been studied as a binder for wood adhesives with promising results [34–44]. Other proteins that have been evaluated are, for example, zein protein [45], pea protein [45], canola protein [46, 47], cotton-seed protein [47], triticale protein [48], and lupine protein [49, 50].

3. Polysaccharides

Polysaccharides, built up by hydroxyl-functional monosaccharides joined together by glycosidic bonds, are an interesting group of polymers that have shown potential as binders for wood adhesives [51]. A polysaccharide with high molar mass will provide cohesive strength to the adhesive; however, a high molar mass will also give higher viscosity. The formation of hydrogen bonds between hydroxyl groups in the polysaccharide and the substrate allows for strong adhesion to wood. However, the hydroxyl groups also render the polysaccharide hydrophilic, which has a negative impact on the water resistance of the final adhesive. Improving the water resistance of polysaccharides is a challenge that is subject to much research.

3.1. Starch

Starch is a polysaccharide that has shown great potential as a binder for wood adhesives [14, 15, 52]. Starch has attracted much attention because of its abundance, renewability, and low price. Starch consists of amylose and amylopectin. Amylose is a long, linear polymer of α-1,4-linked D-glucopyranose, and amylopectin is a branched polymer, much larger than amylose, with α-1,4-linked glucose segments connected with α-1,6-linked branch points. The proportion of amylose to amylopectin affects the properties of the wood adhesive.

Native starch is a very good binder for wood, but it suffers from insufficient water resistance, when modifications are necessary. Many strategies have been employed to improve the properties of starch-based wood adhesives. Starch is often combined with another component, for example, polyvinyl alcohol, or polyvinyl acetate, to increase water resistance [53, 54]. There are several studies on graft polymerizations from starch, commonly using vinyl acetate as monomer and ammonium persulfate as initiator [54–56]. Confocal Raman Microscopy has been used to study the homogeneity of vinyl acetate-grafted starch. It was shown that the graft efficiency was important and had a large effect on the bonding performance of the starch adhesive [57].

One of the main challenges with starch-based wood adhesives is the storage stability; lately, several studies have demonstrated how this can be improved. The effect of emulsifiers in wood adhesives with high starch content has been studied, since the addition of emulsifiers can allow for good dispersion and storage stability. A combination of lauryl sodium sulfate (LSS) and alkylphenol ethoxylates (APEO) can improve the flowability and stability of the adhesive. However, a substitute for APEO is necessary since APEO has a negative impact on the environment [58]. Heat treatment of starch prior to the grafting reaction has been shown to enhance the grafting efficiency and the storage stability [59]. Sucrose-fatty acid esters, derived from renewable sources, have been used as surfactants added prior to grafting a high amylose-containing starch with vinyl acetate. The addition of surfactant resulted in excellent storage stability and improved bonding performance. The thermal stability increased and it was suggested that aggregation of particles was impeded by the surfactant [60]. Addition of SDS has been shown to improve the dispersibility and storage stability of starch grafted with vinyl acetate [61]. Unfortunately, the bonding performance was decreased with the addition of SDS. By adding Na-montmorillonite, a nano-layered silicate, the bonding performance was improved. An addition of 5% Na-montmorillonite enhanced the strength both in dry and wet state; the viscosity slightly increased but the dispersibility and stability were retained [62].

Acid hydrolysis of starch is a common method that can be used to modify starch and improve the solubility and the viscosity of the adhesive. Acid hydrolysis results in breakage of hydrogen bonds between starch molecules, facilitating starch molecules to react with crosslinkers or grafting monomers. Excessive acid hydrolysis may, however, damage the structure. More recently, studies have been performed to better understand how acid hydrolysis can be used to improve bonding performance, water resistance, heat resistance, and storage stability of the adhesive [63, 64].

A starch wood adhesive has been synthesized by oxidation of starch and subsequent attachment of a silane coupling agent, followed by polymerization of butyl acrylate and vinyl acetate. The bonding performance and the thermal stability of the adhesive were improved [65]. Butyl acrylate-grafted starch also resulted in an adhesive with good stability. The bonding performance was similar to a commercial polyvinyl acetate adhesive even after storage for 3 months [65]. In another paper, oxidized starch was grafted with acrylamide and butyl acrylate and thereafter crosslinked with vinyltriisopropoxysilane to yield an adhesive with improved bonding performance [66]. A γ-methacryloxypropyl trimethoxy silane coupling agent was also studied for crosslinking starch grafted with vinyl acetate. The resulting adhesive showed improved bond strength, storage stability, and shear-thinning properties [67].

An isocyanate pre-polymer has been used to crosslink starch to enhance the water resistance. The starch was prepared by oxidation, or esterification, to reduce the crystallinity and increase the reactivity, and thereafter crosslinked with isocyanate [68–70]. Carboxymethyl cellulose (CMC) is an another component that has been added to a starch/isocyanate adhesive together with polyvinyl alcohol and borax to tailor the viscosity, solids content, bonding performance, and the interface compatibility between starch and isocyanate pre-polymer [71].

3.2. Chitosan

Chitosan is deacetylated chitin, which is a polysaccharide that exists in crustaceans such as shrimp and crab, and insects. Naturally, chitosan only occurs in *Mucoraceae* fungi, but chitosan is also readily available and abundant by means of easy and facile derivatization from chitin. It is estimated that 10 billion tons of chitin are synthesized every year by the related organisms [72]. Chitosan is a polysaccharide consisting of β-(1,4)-linked 2-acetamido-2-deoxy-D-glucopyranose (N-acetyl glucosamine) and 2-amino-2-deoxy-D-glucopyranose (glucosamine). Besides being an economically feasible bioresource, chitosan has received great attention for a wide range of commercial applications [73–76] as it is biodegradable, biocompatible, non-toxic, antimicrobial and has reactive amino side groups which allow for chemical modification.

Besides the molecular weight, the degree of deacetylation is particularly important for the adhesion properties of chitosan. Chitosan is characterized by reactive amino groups and numerous hydroxyl groups that can interact with many different functional groups and high molar mass that provides cohesive strength, which makes it an interesting group of materials. Chitosan is insoluble in most organic solvents but soluble in water at acidic pH due to protonation of the amino groups. Combined, these characteristics open up the possibility for new applications in the adhesive and binder area [77].

Chitosan has shown great potential as a binder in adhesives, with or without additives or crosslinkers, as reviewed in literature [77, 78]. Chitosan formulations were prepared by dissolving chitosan in acetic acid solution, often used to dissolve chitosan. Double-lap wood specimens were bonded and dried for 24 h at 40°C with no applied pressure. The adhesive showed good bond strength that could be further improved with the addition of glycerol and trisodium citrate dehydrate [79]. It has been reported that citric acid reacts with chitosan amine groups to form amide bonds [80] and glycerol acts as a plasticizer which takes part in the curing process [81]. However, all formulations were found to have poor water resistance. The penetration of rhodamine-labeled chitosan in a pinewood matrix was investigated using a micro-imaging technique. No in-depth penetration of high molecular weight chitosan was observed [79].

Chitosan has been modified with glucose via the Maillard reaction [82, 83], aiming at improving bonding properties. The effect of glucose addition to chitosans of varying molecular weight was evaluated with respect to bonding properties [84]. It was found that the dry and wet strengths were improved for low-molecular-weight chitosan. However, addition of glucose had a negative impact on bonding properties of high-molecular-weight chitosan.

Chitosan adhesives have also been prepared in combination with konjak glucomannan (KGM) [85]. Compared with casein and soybean protein references, the chitosan-KGM adhesive showed better bonding performance, but was inferior to a UF resin [85]. The bonding performance of chitosan-KGM-polyvinyl alcohol adhesive was also modeled using the Box-Behnken design for response surface methodology [86]. Polypeptides have been used to react with aldehyde and amine groups present in KGM and chitosan, respectively. The wet strength was improved, but the dry strength deteriorated [87].

3.3. Hemicelluloses/gums

Hemicelluloses can be found together with cellulose and lignin in biomass such as wood, grass, and cereals, while gums can be found in seeds, plants, seaweeds or microorganisms for example. Hemicelluloses and gums are heteropolysaccharides consisting of sugars such as xylose, arabinose, mannose, glucose, galactose, and sugar acids, and their chemical composition and structure varies with species [51, 88]. Hemicelluloses have low molecular weight, the average degree of polymerization (DP) is typically 80–200, compared with gums or cellulose with DP of several thousands [7]. Hemicelluloses and gums are among the most abundant biopolymers in nature. By-products from forestry and agriculture are good sources of hemicelluloses. With the growing environmental awareness, the traditional pulp mill is converted into a biorefinery to make efficient use of all possible side streams. Apart from power and fuel production, the possibilities for deriving value-added chemicals are also exploited. The hemicellulose-rich hydrolysates, emanating from the pulping process, can be explored as a constituent for wood adhesives. However, the hemicelluloses in the dilute hydrolysate have rather low-molecular weight and the liquor also contains salts and other by-products; therefore, further fractionation/purification is needed before evaluated in wood adhesives. Hemicelluloses are similar to starch structurally, thus holding promise for adhesive applications; yet, the challenges regarding water resistance remain. Moreover, hemicelluloses, unlike starch, have the advantage of not being a food source.

Locust bean gum (LBG), guar, tamarind, and xanthan gum have been evaluated as wood adhesives [89]. LBG, a galactomannan, obtained from seeds from the carob tree, has shown promising properties. A water dispersion of LBG exhibited very good bond strength, water resistance, and heat resistance [89]. A bacterial polysaccharide, FucoPol, has been tested as a binder for wood. The bond strength was fairly good, but the water resistance was very poor [90]. Another bacterial polysaccharide, xanthan gum, has been evaluated in wood adhesives [89, 91]. Xanthan gum by itself does not show any water resistance, but oxidized xanthan gum, on the other hand, show water resistance similar to polyurethane and chitosan adhesives. No significant improvement was observed by combining xanthan gum with chitosan [91]. Gum Arabic has been evaluated to produce particleboards that noticeably improved the overall panel properties. The particleboard had a smooth surface, and improved internal bond strength [92]. It has been reported that KGM can be used as a wood adhesive with good bond strength and stability [93]; however, low water resistance is limiting its applicability in wood adhesives [85]. The adhesion of KGM on plywood has been investigated and it was found that the dry bond strength was relatively good despite the low solid content used [85]. However, the water resistance was found to be very poor. In several studies, KGM has been combined with chitosan, as reviewed in Section 3.2.

Xylan extracted from beech has been suggested as a binder in wood adhesives [94]. Xylan itself does not show sufficient bonding performance, especially regarding water resistance. A combination of xylan with dispersing agents with/without crosslinkers has been studied. The best combination was a water dispersion of xylan and polyvinyl amine that exhibited good bond strength, water resistance, and heat resistance [94].

4. Lignin

Lignin is the third most abundant biological macromolecule and exists in the complex and rigid structure of lignocelluloses [1, 95, 96]. It is present in lignocellulosics including wood, grass, agricultural residues, and other plants [97]. It is a high-molecular-weight aromatic polymer based on phenylpropane units in a densely crosslinked structure. Intimately interspersed in the lignocellulose structure, lignin acts as a "glue" binding cellulose and hemicellulose together, thus providing rigidity and microbial resistance to the cell wall [98].

Chemical modification of lignin has been extensively studied [99]. Lignin in adhesive applications has been actively researched and several papers have been published recently on this topic [14, 15, 100, 101]. The main interest in lignins as adhesives is due to its structural similarity to phenol, suggesting they can be used as substitutes for phenol-formaldehyde (PF) resins [102]. In adhesive research, lignin is often combined with synthetic resins such as PF and/or urea-formaldehyde (UF) resins [103] to decrease cost [104] and formaldehyde emission [105].

Lignin in adhesives have rendered limited commercial success mainly due to the low reactivity toward formaldehyde, or other aldehydes, as a result of its complexity and low number of reactive sites [101]. The low reactivity is a disadvantage in applications where short curing times are desired. To enable utilization in various material applications, the reactivity can be improved by modification [106–108]. Phenolic and aliphatic hydroxyl groups on lignin allow for chemical modifications including esterification, etherification, phenolation, oxidation, reduction, and amination.

Lignin was used without any chemical modifications as a phenol substitute in the synthesis of a resin evaluated for plywood applications [109]. This study involved the replacement of phenol with lignin in a PF resol resin. The effect of the formaldehyde ratio was studied, and the prepared plywoods were found to exhibit very low formaldehyde emissions. In terms of thermomechanical properties, the lignin-based resin was reported to be similar to the PF reference. Organosolv and sulphite lignin were phenolated and used as reactive precursors for wood veneers and particleboards [102]. Wet and dry internal bond strength fulfills European standards for load-bearing boards in humid environments. In a recent study, a lignin-urea-glyoxal (LUG) wood adhesive was prepared in attempt to eliminate the use of formaldehyde from wood-based panels such as plywood [95]. No significant decrease in mechanical strength was observed, when it was concluded that glyoxal is a suitable substitute for formaldehyde in wood adhesives. In the same study, further improvement of physical and mechanical properties was obtained by adding nanoclay to the LUG resin. Lignin was modified by phenol and used in UF resins [110]. It was demonstrated that the formaldehyde emission from wood panels, bonded with modified lignin-UF resin, was lower compared with commercial UF adhesives. Wood panels were also prepared by using lignin in a UF resin [111, 112]. Prior to the resin preparation, lignin was modified by two different methods, either by a pretreatment with a green ionic liquid (IL), 1-ethyl-3-methylimidazolium acetate, or by glyoxalation, to increase the reactivity [111]. Both modifications resulted in UF resins with improved properties. Panels bonded with IL-modified lignin UF resins were superior, regarding mechanical strength, formaldehyde emission, and gel time. The effect of addition of polymeric 4,4-diphenylmethane diisocyanate (pMDI) liquid to (IL)-treated lignin-urea-formaldehyde resin on the physical and mechanical properties of plywood panels ionic have been investigated [112]. The addition of pMDI enhanced the performance of lignin-UF resins for wood-based panels.

Lignin-based adhesives have been formulated by substituting all phenols with unmodified lignin from corn stover in a PF resin. The mechanical strength of the evaluated adhesive system was similar to a commercial phenol-resorcinol-formaldehyde (PRF) adhesive reference [113]. Different industrial lignins, such as softwood Kraft lignins, have been evaluated in self-binding high-density fibreboards, of which some were found to perform well in dry conditions [114]. Lignin has also been blended with other bio-based polymers such as soybean protein and/or chitosan [32, 115]. Lignin amine was prepared by an efficient two-step process, in which lignin was oxidized through a Fenton oxidation reaction, followed by a reductive amination to yield lignin amine [32]. The soybean protein-lignin amine system exhibited high dry and wet strengths for plywood. The improved properties were attributed to the catechol-like functionalities in the system, mimicking marine adhesive proteins. Laccase-modified lignin, combined with either soybean protein or chitosan, followed by reduction with $NaBH_4$ was evaluated as cheap and safe adhesive systems [115]. The dry strength of the lignin-chitosan formulation was slightly improved when laccase-treated lignin was used, but the subsequent reduction drastically reduced the bonding strength. The laccase-treated and reduced lignin-soybean protein adhesive exhibited more than half the strength compared with a commercial polyurethane adhesive and showed good water resistance.

5. Tannin

Tannins are polyphenols that exist naturally in the bark of various trees such as mimosa, quebracho or pine [116–118]. Natural tannins are divided into two main classes: hydrolysable tannins and condensed polyflavonoid tannins. Condensed tannins constitute more than 90% of the total production of commercial tannins [119]. Tannins are water-soluble compounds [120], and similar to lignin [121], they are of particular interest in the replacement of phenolic resins owing to their chemical structure, similar to phenolic compounds, as well as the ability to react with formaldehyde [120].

The high reactivity toward aldehydes and other reagents renders tannins both chemically and economically interesting in the production of adhesives and resins [120]. However, the present use of commercial tannins for leather manufacturing [122], and beverages [123] limits their availability for other industrial resins. Despite this, tannins have been the subject of extensive research, leading to the exploration of a wide range of adhesive applications [14, 101, 124]. Attempts have been made to commercialize tannin-based adhesives with limited success [125].

In a recent study, tannins have been suggested as scavengers for formaldehyde in MF resins. The formaldehyde emission was decreased with the addition of tannin [126]. The effect of addition of tannin in PF and MF resins was also studied, and the observed results were improved water resistance and decreased formaldehyde emission [127].

Tannins have also been used as a substitute for phenols in PF resins with promising results. Good characteristics and low formaldehyde emissions were obtained even though only a part of the phenol was replaced by tannin [116]. Noteworthy, curing agents used to prepare such adhesive systems, e.g., tris(hydroxymethyl)nitromethane, paraformaldehyde, hexamethylenetetramine, may cause formaldehyde emission [128, 129]. A hyperbranched polyamine ester was used as a curing agent together with tannin-glyoxalated Kraft lignin (TGKL) and evaluated as an adhesive [120]. The adhesive was evaluated on plywood and the results showed that the modification improved the water resistance strength of TGKL effectively. In another study, lignin-based aldehydes were combined with a reactive tannin from pine bark and evaluated as adhesive in wood panels. The performance was in accordance with relevant standards [130].

Boric acid (BA), a weak Lewis base, is known to induce tannin gelation [131]. Hexamine was used in tannin-BA adhesive systems to fixate BA through complexing onto the network, thus preventing leaching of BA. It is also reported that this treatment enhanced biological resistance [132]. Tannin has been explored for plywood applications by incorporating BA into the adhesive system to increase the hardening rate, thus decreasing press time [133]. The adhesive system was composed of quebracho tannin, NaOH, hexamine, BA, and polymeric isocyanate (pMDI). The results indicated that the addition of BA not only lowered time and temperature of hardening, which is of significant importance for the cost of production of wood composites, but also resulted in higher elastic modulus and good resistance to fungal attack.

Recently, tannin adhesives have also been studied in the preparation of particleboards [122, 134, 135]. Internal bond (IB) strengths of the particleboards were investigated with different tannin-based adhesive formulations [122]. It is reported that particleboards manufactured with the formulation containing paraformaldehyde powder as a hardener exhibited the best IB strength. A formaldehyde-free adhesive based on tannin for manufacturing of particleboards has also been developed [134]. The effect of type and concentration of hardeners was investigated. Tris(hydroxymethyl)nitromethane (TRIS), glyoxal (GLY), and hexamethylenetetramine (HEX) were used as hardener instead of formaldehyde. Condensed chestnut-shell tannins could be combined with low proportions of chestnut bur or eucalyptus bark without reducing board quality. The formaldehyde emission level was low.

Glulam from three layers of lamina has been manufactured by using a mahogany tannin-adhesive system, composed of tannin extract and formaldehyde [124]. The results showed that the adhesive was on par with the conventional adhesive for glulam manufacturing regarding mechanical and physical properties, and had low formaldehyde emission. In a more recent study [136], an adhesive of mimosa tannin, mixed with the *Eremurus* root (syrysh), was prepared to manufacture three-layer particleboards, yielding a fully green product. The addition of syrysh improved the overall performance of the mimosa tannin adhesive system.

6. Outlook

The industry's transition to fully green wood adhesives, i.e., based solely on sustainable bio-polymers, will most likely linger for quite some time. Still, substantial challenges remain to develop bio-based adhesives that fulfill all the prerequisites for both indoor and outdoor applications. Water resistance is one area that is particularly challenging; the general, inherent hydrophilic character of most bio-polymers has to be tampered in such a way that the final adhesive bond is able to withstand both humidity and water sufficiently. There are already promising solutions on how to tackle this, but so far it has almost only been demonstrated on a lab scale and not for large-scale productions. Bio-polymers are also encompassed with large natural variations (molecular weight, solubility, etc.) as an effect of shifting growth conditions which bring about a need for a greater flexibility in the production line, either when manufacturing the adhesive or when conducting the actual bonding. On a lab scale, this is usually manageable but it rapidly becomes a problem when targeting large-scale manufacturing. However, there is light in the tunnel. Hybrid adhesive systems, partly composed of bio-based polymers, already exist and suggest that the transition to greener adhesive systems will occur gradually. Stricter legislations and tougher social expectations are believed to accelerate this development. It is, however, challenged by manufacturers' and end-users' unwillingness to pay more for a greener adhesive, when the transition has to be justified by a neutral cost performance and/or improved final properties. For researchers around the world, the field of green adhesives is an exciting area to be active in. Fundamental research is still required in order to develop detailed understanding of the complex adhesion mechanisms that govern strong bonding between adhesive and substrates. A better understanding will propel the future design of green adhesive systems, and scientific breakthroughs may have a substantial impact on a large industrial sector.

Acknowledgements

This study was supported by VINNOVA VINN Excellence Centre BiMaC Innovation, and Wobama WoodWisdom –Net ERA-NET/Formas.

Author details

Emelie Norström[1], Deniz Demircan[1], Linda Fogelström[1], Farideh Khabbaz[2] and Eva Malmström[1]*

*Address all correspondence to: mavem@kth.se

1 Department of Fibre and Polymer Technology, School of Chemical Science and Engineering, KTH Royal Institute of Technology, Stockholm, Sweden

2 AkzoNobel Adhesives AB, Årsta, Sweden

References

[1] Frihart CR. Wood adhesion and adhesives. In: Rowell RM, editor. Handbook of Wood Chemistry and Wood Composites. Florida: CRS Press; 2005. pp. 215-278

[2] Akzo Nobel Adhesives. History. 2017. Available from: https://www.akzonobel.com/adhesives/aboutus/history/

[3] Lambuth AL. Protein adhesivs for wood. In: Pizzi A, Mittal KL, editors. Handbook of Adhesive Technology. Second ed. New York: Marcel Dekker, Inc; 2003. pp. 457-477

[4] Bye CN. Casein and mixed protein adhesives. In: Skeist I, editor. Handbook of Adhesives. New York: Van Nostrand Reinold; 1990. pp. 135-152

[5] EPA United States Environmental Protection Agency. EPA Issues Final Rule to Protect the Public from Exposure to Formaldehyde, 2017. Available from: https://www.epa.gov/newsreleases/epa-issues-final-rule-protect-public-exposure-formaldehyde

[6] WFP FN:s livsmedelsprogram. Hunger 2017. 2017. Available from: http://sv.wfp.org/hunger/fr%C3%A5gor-om-hunger

[7] Teleman A. Hemicelluloses and pectins. In: Gellerstedt G, editor. The Ljungberg Textbook Wood Chemistry and Wood Biotechnology. Stockholm; 2009. pp. 87-108

[8] Photocredits for pictures of particleboard and plywood distributed by a CC-BY-SA 3.0 license. 2017. Available from: https://commons.wikimedia.org/wiki/File%3ABirch_plywood.jpg; https://commons.wikimedia.org/wiki/File%3AParticleboard.jpg

[9] Frihart CR. Adhesive bonding and performance testing of bonded wood products. Journal of ASTM International. 2005;**2**(7):1-1

[10] Pizzi A, Mittal KL. Wood Adhesives. Boston: CRC Press; 2010

[11] Stoeckel F, Konnerth J, Gindl-Altmutter W. Mechanical properties of adhesives for bonding wood—A review. International Journal of Adhesion and Adhesives. 2013;**45**:32-41

[12] Nelson DL, Cox MM, Lehninger AL. Lehninger's principles of biochemistry. 4th ed. New York: Freeman WH; NY, USA, 2005. 75 p

[13] Vnucec D, Kutnar A, Gorsek A. Soy-based adhesives for wood-bonding—A review. Journal of Adhesion Science and Technology. 2017;**31**(8):910-931

[14] Hemmilä V, Adamopoulos S, Karlsson O, Kumar A. Development of sustainable bio-adhesives for engineered wood panels—A review. RSC Advances. 2017;**7**(61):38604-38630

[15] Ferdosian F, Pan Z, Gao G, Zhao B. Bio-based adhesives and evaluation for wood composites application. Polymers (Basel, Switz). 2017;**9**(2):70/1-/29

[16] Kumar R, Choudhary V, Mishra S, Varma IK, Mattiason B. Adhesives and plastics based on soy protein products. Industrial Crops and Products. 2002;**16**(3):155-172

[17] North Carolina Soybean Production Association. Uses of soybeans. 2017. Available from: http://ncsoy.org/media-resources/uses-of-soybeans/

[18] Solenis. Soyad™ Adhesive Technology. 2017. Available from: https://solenis.com/en/industries/specialties-wood-adhesives/innovations/soyad-adhesive-technology/

[19] Sun XS. Thermal and mechanical properties of soy proteins. In: Wool R, Sun XS, editors. Bio-Based Polymers and Composites. United States: Elsevier Inc; 2005. pp. 292-326

[20] Fan B, Zhang LP, Gao ZH, Zhang Y, Shi J, Li J. Formulation of a novel soybean protein-based wood adhesive with desired water resistance and technological applicability. Journal of Applied Polymer Science. 2016;**133**(27) n/a

[21] Zhang B, Fan B, Huo P, Gao Z-H. Improvement of the water resistance of soybean protein-based wood adhesive by a thermo-chemical treatment approach. International Journal of Adhesion and Adhesives. 2017;**78**:222-226

[22] Zeng N, Xie J, Ding C. Properties of the soy protein isolate/PVAc latex blend adhesives. Advanced Materials Research (Durnten-Zurich, Switzerland) . 2012;**550-553**(Pt. 2, Advances in Chemical Engineering II):1103-7

[23] Luo J, Li L, Luo J, Li X, Li K, Gao Q. A high solid content bioadhesive derived from soybean meal and egg white: Preparation and properties. Journal of Polymers and the Environment. 2017;**25**(3):948-959

[24] Yuan C, Chen M, Luo J, Li X, Gao Q, Li J. A novel water-based process produces eco-friendly bio-adhesive made from green cross-linked soybean soluble polysaccharide and soy protein. Carbohydrate Polymers. 2017;**169**:417-425

[25] Zheng P, Li Y, Li F, Ou Y, Lin Q, Chen N. Development of defatted soy flour-based adhesives by acid hydrolysis of carbohydrates. Polymers (Basel, Switz). 2017;**9**(5):153/1−/12

[26] Jang Y, Li K. An all-natural adhesive for bonding wood. Journal of the American Oil Chemists' Society. 2015;**92**(3):431-438

[27] Luo J, Li C, Li X, Luo J, Gao Q, Li J. A new soybean meal-based bioadhesive enhanced with 5,5-dimethyl hydantoin polyepoxide for the improved water resistance of plywood. RSC Advances. 2015;**5**(77):62957-62965

[28] Wang F, Wang J, Wang C, Chu F, Liu X, Pang J. Fabrication of soybean protein-acrylate composite mini-emulsion toward wood adhesive. European Journal of Wood and Wood Products. 2017:Ahead of Print

[29] Chen M, Chen Y, Zhou X, Lu B, He M, Sun S, et al. Improving water resistance of soy-protein wood adhesive by using hydrophilic additives. BioResources. 2015;**10**(1):41-54, 14 pp

[30] Luo J, Luo J, Yuan C, Zhang W, Li J, Gao Q, et al. An eco-friendly wood adhesive from soy protein and lignin: Performance properties. RSC Advances. 2015;**5**(122):100849-100855

[31] Pradyawong S, Qi G, Li N, Sun XS, Wang D. Adhesion properties of soy protein adhesives enhanced by biomass lignin. International Journal of Adhesion and Adhesives. 2017;**75**:66-73

[32] Xin J, Zhang P, Wolcott MP, Zhang J, Hiscox WC, Zhang X. A novel and formaldehyde-free preparation method for lignin amine and its enhancement for soy protein adhesive. Journal of Polymers and the Environment. 2017;**25**(3):599-605

[33] Cheng HN, Ford C, Dowd MK, He Z. Soy and cottonseed protein blends as wood adhesives. Industrial Crops and Products. 2016;**85**:324-330

[34] D'Amico S, Mueller U, Berghofer E. Effect of hydrolysis and denaturation of wheat gluten on adhesive bond strength of wood joints. Journal of Applied Polymer Science. 2013;**129**:2429-2434

[35] El-Wakil NA, Abou-Zeid RE, Fahmy Y, Mohamed AY. Modified wheat gluten as a binder in particleboard made from reed. Journal of Applied Polymer Science. 2007;**106**:3592-3599

[36] Khosravi S, Nordqvist P, Khabbaz F, Oehman C, Bjurhager I, Johansson M. Wetting and film formation of wheat gluten dispersions applied to wood substrates as particle board adhesives. European Polymer Journal. 2015;**67**:476-482

[37] Khosravi S, Khabbaz F, Nordqvist P, Johansson M. Wheat-gluten-based adhesives for particle boards: Effect of crosslinking agents. Macromolecular Materials and Engineering. 2014;**299**:116-124

[38] Khosravi S, Khabbaz F, Nordqvist P, Johansson M. Protein-based adhesives for particleboards. Industrial Crops and Products. 2010;**32**:275-283

[39] Khosravi S, Nordqvist P, Khabbaz F, Johansson M. Protein-based adhesives for particleboards-effect of application process. Industrial Crops and Products. 2011;**34**:1509-1515

[40] Lei H, Pizzi A, Navarrete P, Rigolet S, Redl A, Wagner A. Gluten protein adhesives for wood panels. Journal of Adhesion Science and Technology. 2010;**24**:1583-1596

[41] Nordqvist P, Khabbaz F, Malmström E. Comparing bond strength and water resistance of alkali-modified soy protein isolate and wheat gluten adhesives. International Journal of Adhesion and Adhesives. 2010;**30**:72-79

[42] Nordqvist P, Lawther M, Malmström E, Khabbaz F. Adhesive properties of wheat gluten after enzymatic hydrolysis or heat treatment—A comparative study. Industrial Crops and Products. 2012;**38**:139-145

[43] Nordqvist P, Nordgren N, Khabbaz F, Malmström E. Plant proteins as wood adhesives: Bonding performance at the macro- and nanoscale. Industrial Crops and Products. 2013;**44**:246-252

[44] Nordqvist P, Thedjil D, Khosravi S, Lawther M, Malmström E, Khabbaz F. Wheat gluten fractions as wood adhesives-glutenins versus gliadins. Journal of Applied Polymer Science. 2012;**123**:1530-1538

[45] Santoni I, Pizzo B. Evaluation of alternative vegetable proteins as wood adhesives. Industrial Crops and Products. 2013;**45**:148-154

[46] Bandara N, Esparza Y, Wu J. Exfoliating nanomaterials in canola protein derived adhesive improves strength and water resistance. RSC Advances. 2017;**7**(11):6743-6752

[47] He Z, Chapital DC, Cheng HN, Dowd MK. Comparison of adhesive properties of water- and phosphate buffer-washed cottonseed meals with cottonseed protein isolate on maple and poplar veneers. International Journal of Adhesion and Adhesives. 2014;**50**:102-106

[48] Bandara N, Chen L, Wu J. Adhesive properties of modified triticale distillers grain proteins. International Journal of Adhesion and Adhesives. 2013;**44**:122-129

[49] Khabbaz F, inventor; Akzo Nobel Coatings International B.V., Neth.. assignee. Aqueous adhesive compositions containing lupin proteins and polymers patent WO2012076566A2. 2012

[50] Huang J, Li K. Development and characterization of a formaldehyde-free adhesive from lupine flour, glycerol, and a novel curing agent for particleboard (PB) production. Holzforschung. 2016;**70**(10):927-935

[51] Patel AK, Mathias J-D, Michaud P. Polysaccharides as adhesives: a critical review. Reviews of Adhesion and Adhesives. 2013;**1**(3):312-345

[52] Gadhave RV, Mahanwar PA, Gadekar PT. Starch-based adhesives for wood/wood composite bonding: Review. Open Journal of Polymer Chemistry. 2017;**7**(2):19-32

[53] Imam SH, Gordon SH, Mao L, Chen L. Environmentally friendly wood adhesive from a renewable plant polymer: Characteristics and optimization. Polymer Degradation and Stability. 2001;**73**:529-533

[54] Wang Z, Li Z, Gu Z, Hong Y, Cheng L. Preparation, characterization and properties of starch-based wood adhesive. Carbohydrate Polymers. 2012;**88**:699-706

[55] Wang Z, Gu Z, Hong Y, Cheng L, Li Z. Bonding strength and water resistance of starch-based wood adhesive improved by silica nanoparticles. Carbohydrate Polymers. 2011;**86**:72-76

[56] Wu Y, Lv C, Han M. Synthesis and performance study of polybasic starch graft copolymerization function materials. Advanced Materials Research (Durnten-Zurich, Switzerland). 2009;**79-82**:43-46

[57] Wang P, Cheng L, Gu Z, Li Z, Hong Y. Assessment of starch-based wood adhesive quality by confocal Raman microscopic detection of reaction homogeneity. Carbohydrate Polymers. 2015;**131**:75-79

[58] Cheng L, Guo H, Gu Z, Li Z, Hong Y. Effects of compound emulsifiers on properties of wood adhesive with high starch content. International Journal of Adhesion and Adhesives. 2017;**72**:92-97

[59] Zheng X, Cheng L, Gu Z, Hong Y, Li Z, Li C. Effects of heat pretreatment of starch on graft copolymerization reaction and performance of resulting starch-based wood adhesive. International Journal of Biological Macromolecules. 2017;**96**:11-18

[60] Zia ud D, Xiong H, Wang Z, Fei P, Ullah I, Javaid AB, et al. Effects of sucrose fatty acid esters on the stability and bonding performance of high amylose starch-based wood adhesive. International Journal of Biological Macromolecules. 2017;**104**(Part_A):846-53

[61] Li Z, Wang J, Cheng L, Gu Z, Hong Y, Kowalczyk A. Improving the performance of starch-based wood adhesive by using sodium dodecyl sulfate. Carbohydrate Polymers. 2014;**99**:579-583

[62] Li Z, Wang J, Li C, Gu Z, Cheng L, Hong Y. Effects of montmorillonite addition on the performance of starch-based wood adhesive. Carbohydrate Polymers. 2015;**115**:394-400

[63] Wang Y, Xiong H, Wang Z, Zia ud D, Chen L. Effects of different durations of acid hydrolysis on the properties of starch-based wood adhesive. International Journal of Biological Macromolecules. 2017;**103**:819-828

[64] Yu H, Fang Q, Cao Y, Liu Z. Effect of HCl on starch structure and properties of starch-based wood adhesives. BioResources. 2016;**11**(1):1721-1728

[65] Zhang Y, Ding L, Gu J, Tan H, Zhu L. Preparation and properties of a starch-based wood adhesive with high bonding strength and water resistance. Carbohydrate Polymers. 2015;**115**:32-37

[66] Sun J, Li L, Cheng H, Huang W. Preparation, characterization and properties of an organic siloxane-modified cassava starch-based wood adhesive. The Journal of Adhesion. 2016:1-16

[67] Chen L, Wang Y, Zia ud D, Fei P, Jin W, Xiong H, et al. Enhancing the performance of starch-based wood adhesive by silane coupling agent(KH570). International Journal of Biological Macromolecules. 2017;**104**(Part_A):137-144

[68] Qiao Z, Gu J, Lv S, Cao J, Tan H, Zhang Y. Preparation and properties of isocyanate prepolymer/corn starch adhesive. Journal of Adhesion Science and Technology. 2015;**29**(13):1368-1381

[69] Qiao Z, Gu J, Lv S, Cao J, Tan H, Zhang Y. Preparation and properties of normal temperature cured starch-based wood adhesive. BioResources. 2016;**11**(2):4839-4849

[70] Wang S-M, Shi J-Y, Xu W. Synthesis and characterization of starch-based aqueous polymer isocyanate wood adhesive. BioResources. 2015;**10**(4):7653-7666

[71] Qiao Z, Gu J, Zuo Y, Tan H, Zhang Y. The effect of carboxymethyl cellulose addition on the properties of starch-based wood adhesive. BioResources. 2014;**9**(4):6117-6129, 13 pp

[72] Zargar V, Asghari M, Dashti A. A review on chitin and chitosan polymers: Structure, chemistry, solubility, derivatives, and applications. ChemBioEng Reviews. 2015;**2**(3): 204-226

[73] Pantaleone D, Yalpani M, Scollar M. Unusual susceptibility of chitosan to enzymic hydrolysis. Carbohydrate Research. 1992;**237**:325-332

[74] Dodane V, Vilivalam VD. Pharmaceutical applications of chitosan. Pharmaceutical Science & Technology Today. 1998;**1**(6):246-253

[75] Hirano S. Chitin biotechnology applications. Biotechnology Annual Review. 1996;**2**: 237-258

[76] Sechriest VF, Miao YJ, Niyibizi C, Westerhausen-Larson A, Matthew HW, Evans CH, et al. GAG-augmented polysaccharide hydrogel: A novel biocompatible and biodegradable material to support chondrogenesis. Journal of Biomedical Materials Research. 2000;**49**(4):534-541

[77] Mati-Baouche N, Elchinger P-H, de Baynast H, Pierre G, Delattre C, Michaud P. Chitosan as an adhesive. European Polymer Journal. 2014;**60**:198-212

[78] Patel AK. Chitosan: Emergence as potent candidate for green adhesive market. Biochemical Engineering Journal. 2015;**102**:74-81

[79] Patel AK, Michaud P, Petit E, de Baynast H, Grediac M, Mathias J-D. Development of a chitosan-based adhesive. Application to wood bonding. Journal of Applied Polymer Science. 2013;**127**(6):5014-5021

[80] Cui Z, Beach ES, Anastas PT. Modification of chitosan films with environmentally benign reagents for increased water resistance. Green Chemistry Letters and Reviews. 2011;**4**(1):35-40

[81] Domjan A, Bajdik J, Pintye-Hodi K. Understanding of the plasticizing effects of glycerol and PEG 400 on chitosan films using solid-state NMR spectroscopy. Macromolecules. 2009;**42**(13):4667-4673

[82] Umemura K, Kawai S. Preparation and characterization of Maillard reacted chitosan films with hemicellulose model compounds. Journal of Applied Polymer Science. 2008;**108**(4):2481-2487

[83] Umemura K, Kawai S. Modification of chitosan by the Maillard reaction using cellulose model compounds. Carbohydrate Polymers. 2007;**68**(2):242-248

[84] Umemura K, Mihara A, Kawai S. Development of new natural polymer-based wood adhesives III: Effects of glucose addition on properties of chitosan. Journal of Wood Science. 2010;**56**(5):387-394

[85] Umemura K, Inoue A, Kawai S. Development of new natural polymer-based wood adhesives I: Dry bond strength and water resistance of konjac glucomannan, chitosan, and their composites. Journal of Wood Science. 2003;**49**(3):221-226

[86] Gu R, Mu B, Yang Y. Bond performance and structural characterization of polysaccharide wood adhesive made from Konjac glucomannan/chitosan/polyvinyl alcohol. BioResources. 2016;**11**(4):8166-8177

[87] Shang J, Liu H, Qi C, Guo K, Tran VC. Evaluation of curing and thermal behaviors of konjac glucomannan-chitosan-polypeptide adhesive blends. Journal of Applied Polymer Science. 2015;**132**(34):n/a

[88] Salam A, Venditti RA, Pawlak JJ, El-Tahlawy K. Crosslinked hemicellulose citrate-chitosan aerogel foams. Carbohydrate Polymers. 2011;**84**:1221-1229

[89] Norström E, Fogelström L, Nordqvist P, Khabbaz F, Malmström E. Gum dispersions as environmentally friendly wood adhesives. Industrial Crops and Products. 2014;**52**:736-744

[90] Araujo D, Alves VD, Campos J, Coelhoso I, Sevrin C, Grandfils C, et al. Assessment of the adhesive properties of the bacterial polysaccharide FucoPol. International Journal of Biological Macromolecules. 2016;**92**:383-389

[91] Paiva D, Goncalves C, Vale I, Bastos MMSM, Magalhaes FD. Oxidized xanthan gum and chitosan as natural adhesives for cork. Polymers (Basel, Switz). 2016;**8**(7):259/1-/13

[92] Abuarra A, Hashim R, Bauk S, Kandaiya S, Tousi ET. Fabrication and characterization of gum Arabic bonded Rhizophora spp. particleboards. Materials and Design. 2014;**60**:108-115

[93] Wang Y, Guo K-Q, Li J-N, Duan X-F, Li Y-J. Konjac glucomannan/chitosan based adhesive for plywood production. China Wood Industry. 2009;**23**(2):13-15

[94] Norstroem E, Fogelstroem L, Nordqvist P, Khabbaz F, Malmstroem E. Xylan—A green binder for wood adhesives. European Polymer Journal. 2015;**67**:483-493

[95] Younesi-Kordkheili H. Improving physical and mechanical properties of new lignin-urea-glyoxal resin by nanoclay. European Journal of Wood and Wood Products. 2017:1-7

[96] Fu D, Mazza G, Tamaki Y. Lignin extraction from straw by ionic liquids and enzymatic hydrolysis of the cellulosic residues. Journal of Agricultural and Food Chemistry. 2010;**58**(5):2915-2922

[97] Khalil HSA, Alwani MS, Omar AKM. Chemical composition, anatomy, lignin distribution, and cell wall structure of Malaysian plant waste fibers. BioResources. 2007;**1**(2):220-232

[98] Kirk TK a, Farrell RL. Enzymatic "combustion": The microbial degradation of lignin. Annual Review of Microbiology 1987;**41**(1):465-501

[99] Laurichesse S, Averous L. Chemical modification of lignins: Towards biobased polymers. Progress in Polymer Science. 2014;**39**(7):1266-1290

[100] Ghaffar SH, Fan M. Lignin in straw and its applications as an adhesive. International Journal of Adhesion and Adhesives. 2014;**48**:92-101

[101] Pizzi A. Recent developments in eco-efficient bio-based adhesives for wood bonding: Opportunities and issues. Journal of Adhesion Science and Technology. 2006;**20**:829-846

[102] Ibrahim MNM, Zakaria N, Sipaut CS, Sulaiman O, Hashim R. Chemical and thermal properties of lignins from oil palm biomass as a substitute for phenol in a phenol formaldehyde resin production. Carbohydrate Polymers. 2011;**86**(1):112-119

[103] Podschun J, Stücker A, Buchholz RI, Heitmann M, Schreiber A, Saake B, et al. Phenolated lignins as reactive precursors in wood veneer and particleboard adhesion. Industrial and Engineering Chemistry Research. 2016;**55**(18):5231-5237

[104] Shimatani K, Sano Y, Sasaya T. Preparation of moderate-temperature setting adhesives from softwood Kraft lignin. Part 2. Effect of some factors on strength properties and characteristics of lignin-based adhesives. Holzforschung. 1994;**48**(4):337-342

[105] Yang S, Zhang Y, Yuan TQ, Sun RC. Lignin–phenol–formaldehyde resin adhesives prepared with biorefinery technical lignins. Journal of Applied Polymer Science. 2015;**132**(36)

[106] El Mansouri NE, Yuan Q, Huang F. Synthesis and characterization of kraft lignin-based epoxy resins. BioResources. 2011;**6**(3):2492-2503

[107] Stewart D. Lignin as a base material for materials applications: Chemistry, application and economics. Industrial Crops and Products. 2008;**27**(2):202-207

[108] Qin J, Woloctt M, Zhang J. Use of polycarboxylic acid derived from partially depolymerized lignin as a curing agent for epoxy application. ACS Sustainable Chemistry & Engineering. 2013;**2**(2):188-193

[109] Tachon N, Benjelloun-Mlayah B, Delmas M. Organosolv wheat straw lignin as a phenol substitute for green phenolic resins. BioResources. 2016;**11**(3):5797-5815

[110] Younesi-Kordkheili H, Pizzi A, Niyatzade G. Reduction of formaldehyde emission from particleboard by phenolated kraft lignin. The Journal of Adhesion. 2016;**92**(6):485-497

[111] Younesi-Kordkheili H, Pizzi A. A comparison between lignin modified by ionic liquids and glyoxalated lignin as modifiers of urea-formaldehyde resin. The Journal of Adhesion. 2016:1-11

[112] Younesi-Kordkheili H, Pizzi A, Mohammadghasemipour A. Improving the properties of ionic liquid-treated lignin-urea-formaldehyde resins by a small addition of isocyanate for wood adhesive. The Journal of Adhesion. 2017:1-14

[113] Kalami S, Arefmanesh M, Master E, Nejad M. Replacing 100% of phenol in phenolic adhesive formulations with lignin. Journal of Applied Polymer Science. 2017;**134**(30):n/a

[114] Tupciauskas R, Gravitis J, Abolins J, Veveris A, Andzs M, Liitia T, et al. Utilization of lignin powder for manufacturing self-binding HDF. Holzforschung. 2017;**71**(7-8):555-561

[115] Ibrahim V, Mamo G, Gustafsson P-J, Hatti-Kaul R. Production and properties of adhesives formulated from laccase modified Kraft lignin. Industrial Crops and Products. 2013;**45**:343-348

[116] Kim S. Environment-friendly adhesives for surface bonding of wood-based flooring using natural tannin to reduce formaldehyde and TVOC emission. Bioresource Technology. 2009;**100**(2):744-748

[117] Pinto PC, Sousa G, Crispim F, Silvestre AJ, Neto CP. Eucalyptus globulus bark as source of tannin extracts for application in leather industry. ACS Sustainable Chemistry & Engineering. 2013;**1**(8):950-955

[118] Fechtal M, Riedl B. Use of eucalyptus and Acacia mollissima bark extract-formaldehyde adhesives in particleboard manufacture. Holzforschung. 1993;**47**(4):349-357

[119] Pizzi A. Types, processing and properties of bioadhesives for wood and fibers. In: Waldron KW, editor. Advances in Biorefineries: Biomass and Waste Supply Chain Exploitation. Amsterdam: Elsevier; 2014. pp. 736-771

[120] Faris AH, Ibrahim MNM, Rahim AA. Preparation and characterization of green adhesives using modified tannin and hyperbranched poly(amine-ester). International Journal of Adhesion and Adhesives. 2016;**71**:39-47

[121] Faris AH, Rahim AA, Ibrahim MNM, Hussin MH, Alkurdi AM, Salehabadi A. Investigation of oil palm based Kraft and auto-catalyzed organosolv lignin susceptibility as a green wood adhesives. International Journal of Adhesion and Adhesives. 2017;**74**:115-122

[122] Konai N, Pizzi A, Raidandi D, Lagel M, L'Hostis C, Saidou C, et al. Aningre (Aningeria spp.) tannin extract characterization and performance as an adhesive resin. Industrial Crops and Products. 2015;**77**:225-231

[123] Harbertson JF, Parpinello GP, Heymann H, Downey MO. Impact of exogenous tannin additions on wine chemistry and wine sensory character. Food Chemistry. 2012;**131**(3):999-1008

[124] Lestari ASRD, Hadi YS, Hermawan D, Santoso A. Glulam properties of fast-growing species using mahogany tannin adhesive. BioResources. 2015;**10**(4):7419-7433

[125] Pizzi A. Wood products and green chemistry. Annals of Forest Science. 2016;**73**(1):185-203

[126] Kim S, Kim HJ, Kim HS, Lee HH. Effect of bio-scavengers on the curing behavior and bonding properties of melamine-formaldehyde resins. Macromolecular Materials and Engineering. 2006;**291**(9):1027-1034

[127] Gangi M, Tabarsa T, Sepahvand S, Asghari J. Reduction of formaldehyde emission from plywood. Journal of Adhesion Science and Technology. 2013;**27**(13):1407-1417

[128] Kim S, Kim H-J. Curing behavior and viscoelastic properties of pine and wattle tannin-based adhesives studied by dynamic mechanical thermal analysis and FT-IR-ATR spectroscopy. Journal of Adhesion Science and Technology. 2003;**17**(10):1369-1383

[129] Trosa A, Pizzi A. A no-aldehyde emission hardener for tannin-based wood adhesives for exterior panels. Holz als Roh- und Werkstoff. 2001;**59**(4):266-271

[130] Santiago-Medina F, Foyer G, Pizzi A, Caillol S, Delmotte L. Lignin-derived non-toxic aldehydes for ecofriendly tannin adhesives for wood panels. International Journal of Adhesion and Adhesives. 2016;**70**:239-248

[131] Thevenon M-F, Tondi G, Pizzi A. High performance tannin resin-boron wood preservatives for outdoor end-uses. European Journal of Wood and Wood Products. 2009;**67**(1):89-93

[132] Tondi G, Wieland S, Lemenager N, Petutschnigg A, Pizzi A, Thevenon M-F. Efficacy of tannin in fixing boron in wood. BioResources. 2012;**7**(1):1238-1252

[133] Efhamisisi D, Thevenon M-F, Hamzeh Y, Karimi A-N, Pizzi A, Pourtahmasi K. Induced tannin adhesive by boric acid addition and its effect on bonding quality and biological performance of poplar plywood. ACS Sustainable Chemistry & Engineering. 2016;**4**(5):2734-2740

[134] Santos J, Antorrena G, Freire MS, Pizzi A, González-Álvarez J. Environmentally friendly wood adhesives based on chestnut (*Castanea sativa*) shell tannins. European Journal of Wood and Wood Products. 2017;**75**(1):89-100

[135] Fitzken Da Vinci M, Niro J, Kyriazopoulos M, Bianchi S, Mayer I, Eusebio DA, Arboleda JR, et al. Development of medium- and low-density fibreboards made of coconut husk and bound with tannin-based adhesives. International Wood Products Journal. 2016;**7**(4): 208-214

[136] Eghtedarnejad N, Mansouri HR. Building wooden panels glued with a combination of natural adhesive of tannin/Eremurus root (syrysh). European Journal of Wood and Wood Products. 2016;**74**(2):269-272

Chapter 5

A Review of Isocyanate Wood Adhesive: A Case Study in Indonesia

Arif Nuryawan and Eka Mulya Alamsyah

Additional information is available at the end of the chapter

http://dx.doi.org/10.5772/intechopen.73115

Abstract

The use of isocyanate adhesive for the binding of wood and wood products has been increasing in Indonesia particularly for research needs since wood products bonded by glue-based formaldehyde release formaldehyde emission that have been found to have carcinogenic effect and may lead to sick house syndrome. There are at least two types of isocyanate commonly used in Indonesia, namely isocyanate cross-linker and isocyanate alone. Isocyanate cross-linker is used together with polyvinyl alcohol (PVA) forming a water-based polymer-isocyanate emulsion; thus, its application using spreading technique for binding engineered wood products such as glue laminated timber (glulam) and laminated veneer lumber (LVL). For isocyanate alone, because its viscosity is adequate for spraying, it is preferably used for producing wood-based panels, especially particleboard and fiberboard. In this chapter, the characteristics of both types of isocyanate usually used in Indonesia are presented. Some research studies of the authors are also provided.

Keywords: isocyanate, wood adhesive, characteristics

1. Introduction

In order to reduce the formaldehyde emission originated from both adhesive-based formalin and wood products bonded by resin-based formaldehyde, such as urea-formaldehyde (UF), phenol-formaldehyde (PF), melamine-formaldehyde (MF), or a mixture of the two, i.e., melamine-urea-formaldehyde (MUF) or urea-melamine-formaldehyde (UMF), the use of either natural adhesive (bio-adhesive) or isocyanate adhesive has been slightly increased for research purposes in Indonesia. Bio-adhesives have been used by human for thousands of years [1] even though they have limitations in strength and durability [2–4]. On the contrary, isocyanate has the advantage of high strength and resistance and is also stable although it is applied on treated wood [5]. Therefore, the latter has been extensively used, particularly in research of wood products.

Isocyanates have been mainly used for wood adhesive in two ways: first, as a urethane prepolymers originated from isocyanate-polyol reaction products recently being used in the wood-laminating industry, and second, as the isocyanate currently being used in the particleboard industry [6]. The type of isocyanate generally used for binding wood laminated-based products is the water-based emulsion adhesive with isocyanate as the cross-linker [7], while the type of diisocyanate generally used for particleboard manufacture is MDI or 4–4′-diphenylmethane diisocyanate [6]. Further, fiberboard mills involved in particleboard manufacture use MDI as the binder [8, 9]. The method of mass production for particleboard and fiberboard using isocyanate adhesive has been patented in Europe since 2013 [10].

When isocyanate resins are used as a binder for wood materials, the resins react with the wood component [5] and water. However, if water is present in the wood materials, the isocyanate resins would react with the water in preference to the wood component. Therefore, the isocyanate resin-water reaction is considered as the one of the most important reactions when bonding wood composite materials with isocyanate resins.

In this chapter, a review of isocyanate applied on wood would be presented. Types of isocyanates including the chemical composition, their origin or history, properties with an emphasis on thermal behavior, application on the wood products, previous and on-going works of ours have been also explained.

2. History and application of water-based polymer isocyanate in Indonesia

According to our best knowledge, the use of isocyanate in Indonesia originated from the introduction of the product by two Japanese producers of water-based polymer isocyanate (WBPI), namely Koyo Sangyo Co, Ltd. and Oshika Corporation.

Koyo Sangyo then opened the branch factory (PT. Koyolem Indonesia) in 1993 and distributed the WBPI under the trade name Koyobond®(www.koyoweb.com) [12]. Likewise, a joint venture company (PT. Poly Oshika) was established in 1995 between Oshika Corporation and PT. Polichemie Asia Pacific for the production of aqueous polymer isocyanate (API) adhesive in Indonesia under the trade name of PI Bond®(www.polychemie.co.id) [13]. API glue is the same as WBPI, a common name for the adhesive system consisting of an aqueous PVA solution with an isocyanate cross-linker [7]; nowadays, it is also called emulsion polymer isocyanates (EPI) [14]. This information is in accordance with the statement of Grøstad and Pedersen [7], who mentioned from a business point of view that WBPI is naturally concentrated in the Asian markets, presumably including Indonesia. In contrast, non-Asian consumption of this type of adhesive is still limited, although it is slightly increasing.

Application of isocyanate adhesive for gluing wood in Indonesia is in the form of isocyanate for a cross-linker, which is mixed together with PVA, forming a WBPI system for binding engineered wood products such as glulam and LVL by the spreading technique, as shown in **Figure 1**. Glulam is a bonded wood product that is suitable for many applications because it can be utilized as a structural component for settlement construction or as a light structural component in buildings, replacing solid wood products from natural forests, which have been reduced in supply due to enormous logging, forest fire, forest conversion into plantation, and other factors. Furthermore, glulam is considered to be the best alternative material for larger structural components, because it can be manufactured from small laminated lumber.

The best example of Indonesian research on glulam is the work of Herawati et al. [15]. They investigated the quality of glulam made of two Indonesian wood species, namely African wood (*Maesopsis eminii*) and mangium (*Acacia mangium*), and bonded by WBPI with the trade name PI 3100® purchased from PT.Polychemie Asia Pacific, Indonesia. Before they are made into glulam, the sawn lumbers were graded into three classes of modulus of elasticity (MOE) using Panter/Plank-sorter of machine stress grading (MSR), namely E1, E2 and E3. With the composition as shown in **Figure 2**, the glulam was constructed of both balanced and unbalanced combination.

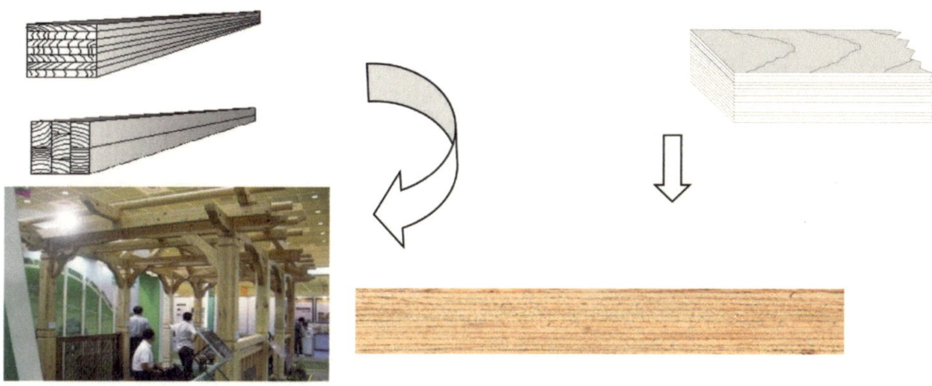

Figure 1. (a) Glulam and (b) LVL, both wood engineered products, were usually glued with a spreading method using an adhesive mixture consisting of PVA and isocyanate cross-linker in the ratio of 100:15 [11].

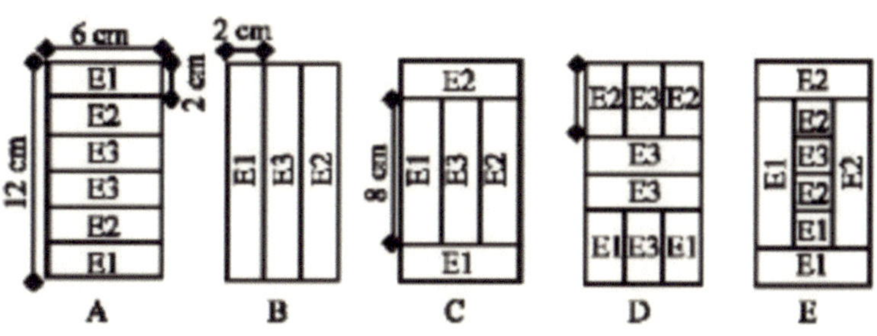

Figure 2. Composition of glulam, work of Herawati et al. [15].

Another product that is usually bonded by WBPI is LVL. It is composed of veneers glued with grain running parallel to each other for reducing the natural variability of wood. Usually the veneer of the higher quality is placed on the outside. Alamsyah et al.[16] investigated the quality of LVL made of three wood species from Indonesia, namely *Paraserianthes falcataria, Pinus merkusii,* and *A. mangium,* and bonded by WBPI (API with trade name of KR Bond-7800®, Koyo, mixed with 15 parts cross-linking agent) at 250 g/m^2 spread rate. They treated the LVL using various physical treatments such as boiling and soaking while the mechanical evaluation consisted of block shear test and contact angle test. Mixing of the glue was done in accordance with the supplier's instruction. Details of the main component of the API adhesive used in their study were presented. **Table 1** exhibits the properties of an aqueous polymer PVAc (polyvinyl acetate) as the resin containing a reactive aqueous polymer. PVAc was produced by hydrolysis of PVA [7].**Table 2** presents the characteristics of the isocyanate cross-linker.

Koyobond® is composed of two components: a resin containing a reactive aqueous polymer and an isocyanate cross-linker. The cross-linker reacts with active groups of not only aqueous polymer but also wood to produce strong chemical bonds.

Further, similar work of Alamsyah et al. [16] included lengthening using eight Indonesian wood species namely EX or *Eucalyptus cyclocarpum,* PF or *P. falcataria,* S or *Shorea* sp., TS or *Toona sinensis,* GA or *Gmelina arborea,* PM or *P. merkusii,* AM or *A. mangium* and AH or *A. hybrid* in order to understand the application and influence of WBPI on wood, the treatment and the testing. The treatment also referred to previous work namely normal or control (N), cyclic boiling (CB), vacuum pressure soaking (VPS) and vacuum pressure

No	Parameter	Properties
1	Appearance	White fluid
2	Viscosity at 25°C	60±20 poise
3	Solid content	58±3%
4	pH	7.5±1.0

Source: PT.LemindoAbadi Jaya [17].

Table 1. Typical properties of PVAc resin (KR-7800®).

No	Remarks	
1	Cross-linker	AJ-1®
2	Mix ratio w/w (resin/cross-linker)	100/15
3	Press time at 25°C	60–120minutes
4	Wood species	"Hard-to-bond wood species" such as teak, sungkai, red oak

Source: PT.Lemindo Abadi Jaya [17].

Table 2. Typical properties of isocyanate cross-linker.

http://dx.doi.org/10.5772/intechopen.73115

soaking dry (VPSD) treatment. **Figure 3** shows the results of the shear strength test of the lamina made from various Indonesian wood species bonded by Koyobond® while **Figure 4** shows the performance of the wood products specimen bonded by Koyobond® after treatment. In this study, shear strength percentages can be used to predict the bond quality of wood-bonded products such as laminated wood. De-lamination is an indicator of how well the bonded joint withstands severe swelling and shrinking stresses in the presence of high moisture and heat [18]. Adhesion using WBPI in this study was resistant to high moisture and heat.

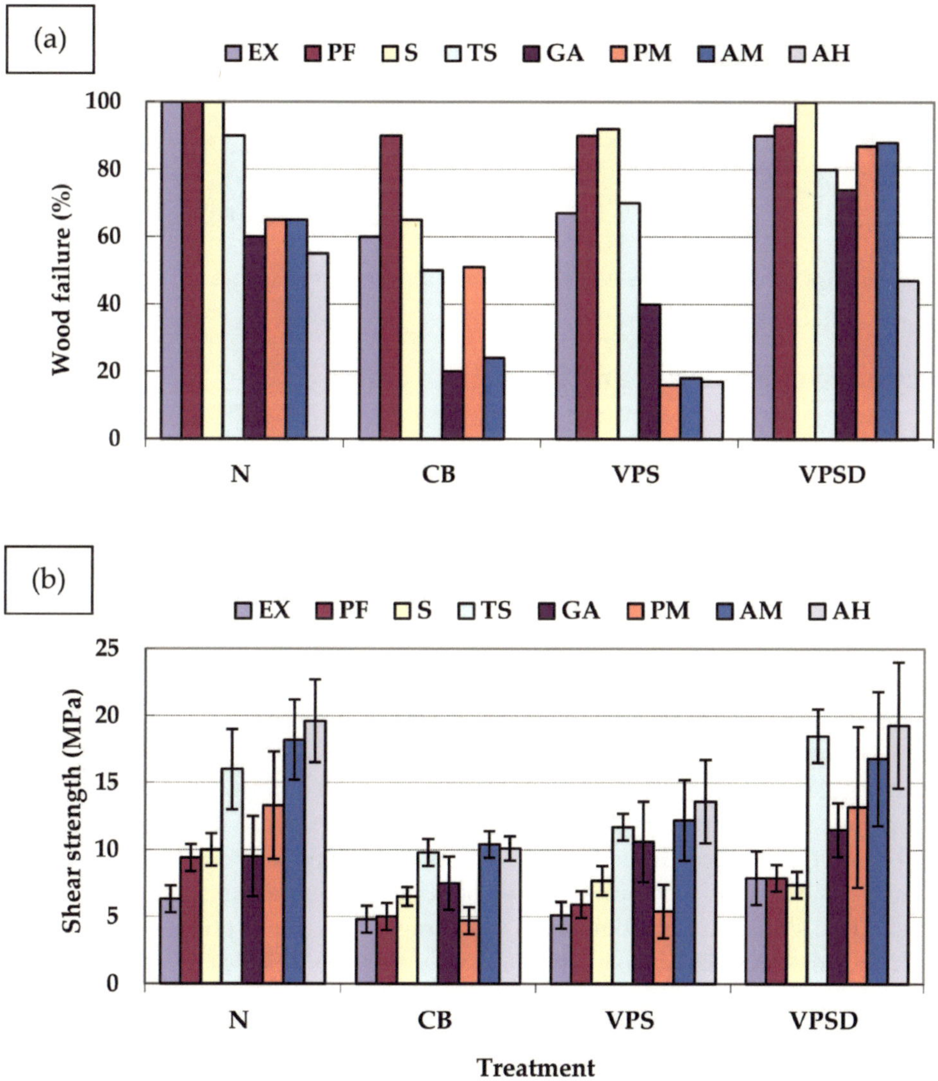

Figure 3. (a) Wood failure and (b) shear strength of laminates bonded with API made from various Indonesian wood species and in various treatments. Error bars indicated a standard deviation [19].

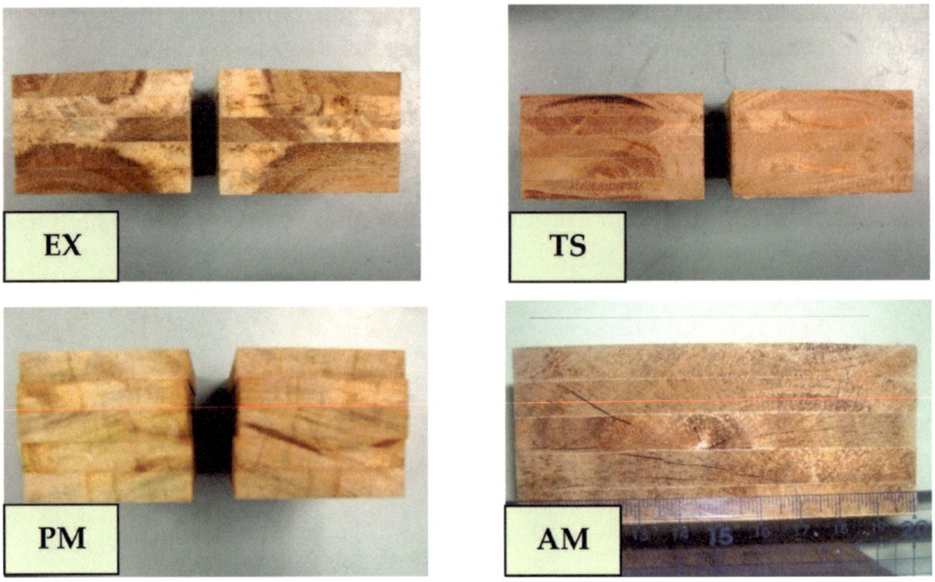

Figure 4. Specimen of wood laminates bonded with Koyobond®after VPS-2 treatment for de-lamination testing [19].

3. Isocyanate cross-linker and isocyanate alone

All isocyanates of industrial importance, including the isocyanate cross-linker used in WBPI, contain two or more isocyanate groups (–N=C=O) per molecule. Isocyanates have good bonding because their –NCO group can react with compounds having active hydrogen such as water, including wood with water therein (free and bonded water) [20–22]. An excellent review and an experiment result published more than three decades ago presented evidence that isocyanates also reacted with the wood component, particularly the cell wall part [5, 23].

To our best knowledge, isocyanate adhesive is a general term for a variety of esters, which rely mainly on phosgene synthesis in conventional synthesis industries. As of now, toluene diisocyanate (TDI) and diphenylmethane diisocyanate (MDI) are the most commercially important diisocyanates [24]; they are widely used and have the largest output [25]. TDI has quite a high vapor pressure; hence, increased health risks involved with the use of this isocyanate are significant [7]. MDI is produced from the reaction between aniline and formaldehyde with hydrochloric acid as the catalyst. A complex mixture of isomeric diamines and oligomeric polyamines is formed. The 4,4′-diamine predominates. This complex mixture is phosgenated to give polymeric diphenylmethane diisocyanate (pMDI), rather than a purified diisocyanate, an adhesive. At ambient temperature, pMDI is a clear brown liquid with a viscosity of about 0.5 Pas and a low vapor pressure. It has an excellent shelf life as long as moisture is not present [20].

pMDI is a mixture of MDI monomer and the related methylene bridged polyphenyl polyisocyanates. In the forest products industry, the binder is commonly referred to as MDI or simply as "isocyanates." This confusion of names is further complicated by the fact that polymeric MDI is not at all polymeric. Approximately one-half of the resin is diisocyanate monomer, while the rest is a complex oligomeric mixture of polyisocyanates with a degree of polymerization less than 12 [9].

In this discussion, the term "isocyanate cross-linker" is used interchangeably with the term "isocyanate alone" because in fact isocyanate alone originates from isocyanate cross-linker as a component of WBPI. In Indonesia, because of the limitation of the producers and suppliers of the isocyanate adhesive, research on wood products bonded by isocyanate was carried out using isocyanate cross-linker purchased from PT.Koyolem Indonesia and PT.Polichemie Asia Pacific.

Figure 5(a) shows the properties of PF resin adhesive and MDI. The color of MDI is brighter than that of PF, but the viscosity is appropriate for applications using the spraying technique for producing wood composite products such as particleboard and fiberboard.

Figure 5(b) shows the packaging of isocyanate cross-linker sold and distributed in Indonesia. Although there is a "hardener" label in the brochure and packaging as shown in the picture, this terminology is wrong. In this context, hardener means cross-linker. Details of the properties are presented in **Table 3**, which contrasts the characteristics of H3M® and H7® (the trade name of the isocyanate cross-linker sold by PT.Polichemie Asia Pacific). Both types of isocyanate cross-linker were capable of being used as adhesive for making fiberboard in laboratory scale [28]. According to the supplier, H3M® is suitable for bonding any wood species; however, H7® is more appropriate for gluing of lamina wood.

Figure 5. (a) Color comparison between PF resin adhesive and MDI; (b) the isocyanate cross-linker distributed and most used in Indonesia [26, 27].

Characteristics	Trade name	
	H3M®	**H7®**
Performance (color)	Brown, viscous	Brown, viscous
Solid content (%)	Min 98–100%	Min 98%
Viscosity (cps/25°C)	180 cps	150–200, ± 175 cps
Source: Polyoshika [29].		

Table 3. Properties of the most used isocyanate cross-linker in Indonesia.

Many researches involving these types of isocyanate cross-linker have been carried out by Indonesian scholars, and the results have been presented or published in conference proceedings or journal papers. For example: Nuryawan et al. [30] mixed UF resin with isocyanate cross-linker purchased from PT.Polyoshika Company in the ratio 100:15 (w/w based on the solid content) for bonding particleboard made of sawdust from the residue of plantation forest of acacia (*A. mangium*) and eucalyptus wood (*Eucalyptus* sp.). Nuryawan et al. [31] used the isocyanate cross-linker of H7® for bonding sawdust's particleboard made of wood industrial waste. Febrianto et al. [32, 33] used commercial MDI (type H3M®) purchased from PT.Polichemie Asia Pacific as the adhesive for oriented strand board (OSB) made from an Indonesian fast growing tree species and betung bamboo, respectively. Recently, Iswanto et al. [34] used isocyanate resin (H3M®) obtained from PT.Polichemie Oshika for binding particleboard made of sorghum bagasse. From these examples, we can sum up that isocyanate cross-linkers are capable of bonding lignocellulose material such as wood particle, bamboo, and even sorghum bagasse.

4. Chemical composition

More than 270 isocyanates were synthesized between 1934 and 1949 [35]. But today, MDI (4,4′-diphenyl methane diisocyanate) is predominant worldwide; it is the generic name of a product used in industrial settings. pMDI or polymeric MDI, the primary technical/commercial form of MDI, is actually a mixture that contains 25–80% monomeric 4,4′-MDI as well as oligomers containing 3–6 rings and other minor isomers, such as the 2,2′-isomer. The exact composition of pMDI varies with the manufacturer [36].

Therefore, in this chapter, we had to emphasize and characterize the most used isocyanate in Indonesia, particularly in the research area of particleboard and fiberboard. Even though we already know the trade name and the supplier as well as the manufacturer, labeling, and writing, the chemical name is suggested. Indeed, MDI or 4,4′-diphenyl methane diisocyanate has at least 10 isomers as stated in the website of PubChem [37], the Open Chemistry Database. The isomers are as follows:

1,1-methylenebis(phenyl)diisocyanate;

4,4′-diisocyanatodiphenylmethane;

4,4′-diphenylmethane diisocyanate;

4,4′-methylene bisphenyldiisocyanate;

4,4′-methylenebis(phenylisocyanate);

4,4′-methylenediphenyl diisocyanate;

diphenylmethanediisocyanate;

diphenylmethane-4,4′-diisocyanate;

methylene diphenyl diisocyanate;

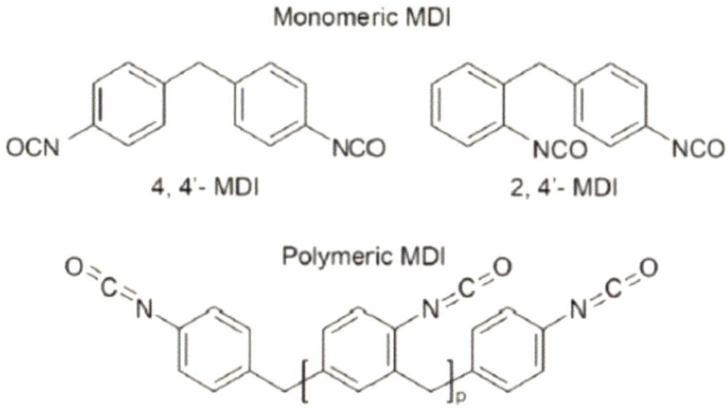

Figure 6. Monomeric MDI in 2 (ortho) position, in 4 (para) position, and its polymeric MDI [38].

p,p′-diphenylmethanediisocyanate.

The main effect of the isocyanate group (–NCO) on reactivity is in the 2 and 4 positions. The isocyanate group in the 2 (ortho) position is three times less reactive than the isocyanate group in the 4 (para) position as shown in **Figure 6**. In addition, pure 4,4′-MDI is solid at ambient temperature.

5. Isocyanate cross-linker applied on fiberboard under the cyclic test

Isocyanate cross-linker can be applied for bonding on either wood particle or wood fiber using spraying methods as shown in **Figure 7**.

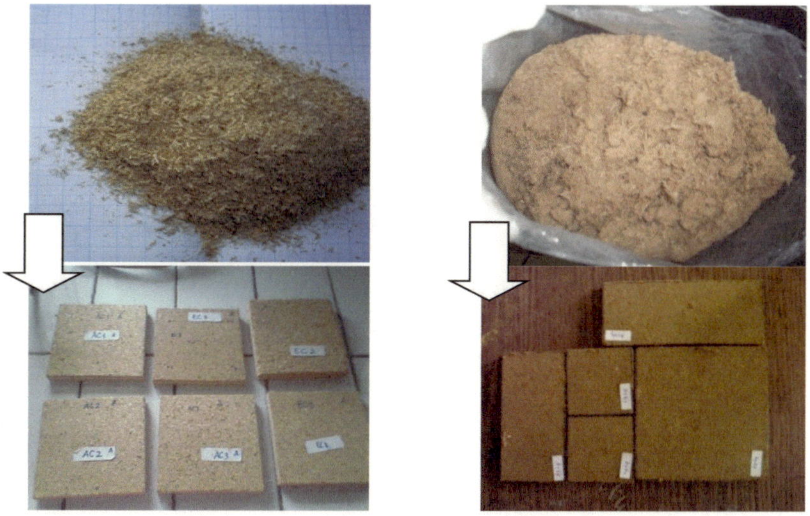

(a) (b)

Figure 7. Typical wood composite products bonded with isocyanate alone using spraying method: (a) particleboard and (b) fiberboard [27, 39].

Nuryawan et al. [27] investigated the physical, mechanical, and performance properties of fiberboard bonded with H3M® and H7®. The result showed that the quality of both fiberboards was similar.

When cyclic evaluation was carried out, comprising water absorption and thickness swelling evaluation, surprisingly H3M® was much less compared to H7®. Moreover, the resistance of the fiberboard bonded by H3M® was stronger after cyclic test. This means that the fiberboard bonded with H3M® showed higher dimension stability compared to the fiberboard bonded with H7® as shown in **Figure 8**.

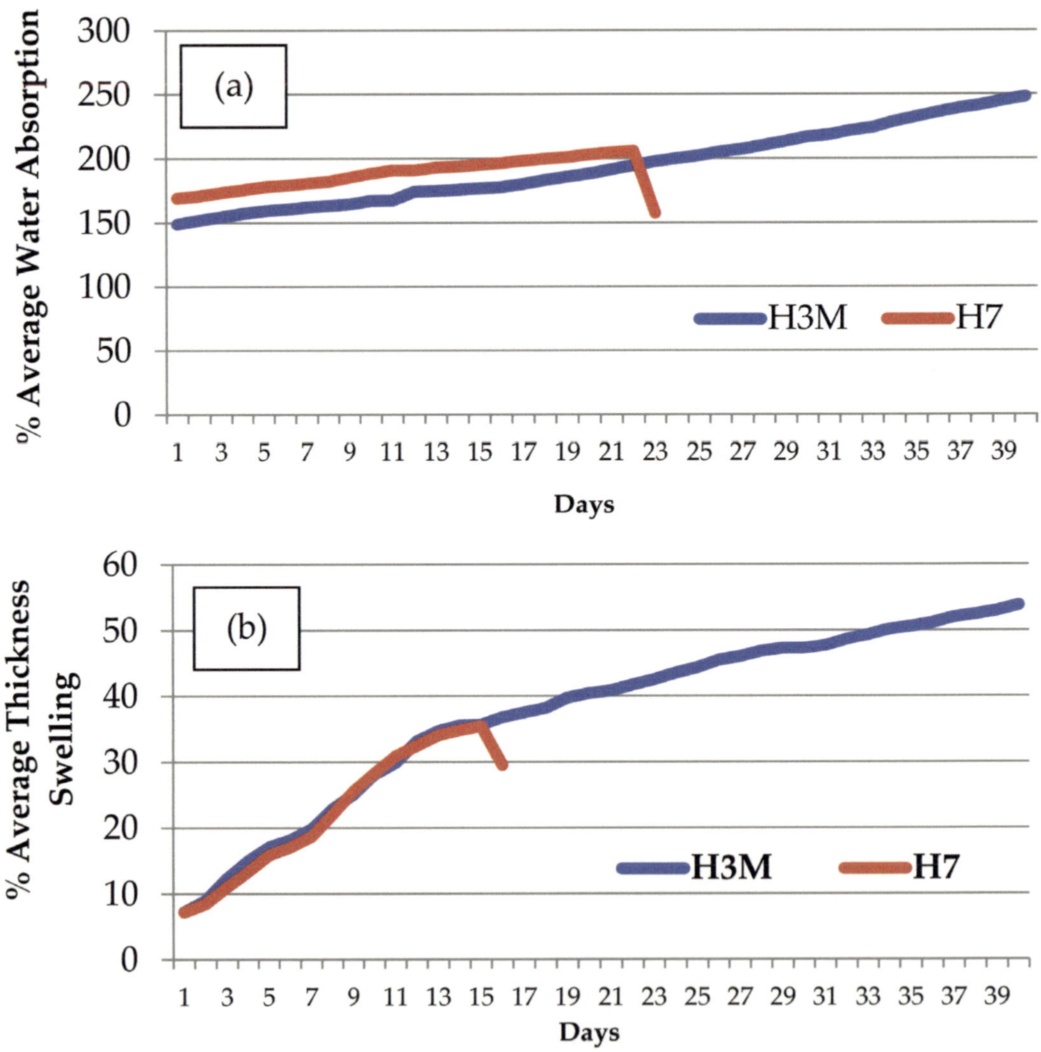

Figure 8. Cyclic test comprising (a) water absorption and (b) thickness swelling showed application of H3M®as binder was much longer in life use and stronger in resistance compared to H7®[27].

6. Other properties of the isocyanate alone

The type of isocyanate alone adhesive used is emulsion polymer isocyanate (EPI). EPI can react with water. On the contrary, pMDI is not dispersible in water because it is oil-borne; therefore, it has to be blocked using blocking agents such as sodium bisulfite ($NaHSO_3$) [40]. A block isocyanate is formed through a reaction between an isocyanate group (–NCO) and a compound containing an active hydrogen atom to block the –NCO. This product has the advantage of a long shelf life because the active isocyanate groups are masked and protected. Furthermore, it has a small amount of the isocynate groups and requires a high temperature to de-block and to make the –NCO group free [41, 42].

On-going researches have been employing different scanning calorimetry (DSC) to analyze the thermal properties of the isocyanate [43]. Results of the analysis of polymeric isocyanate adhesive and its mixture with sawdust are shown in **Figure 9**.

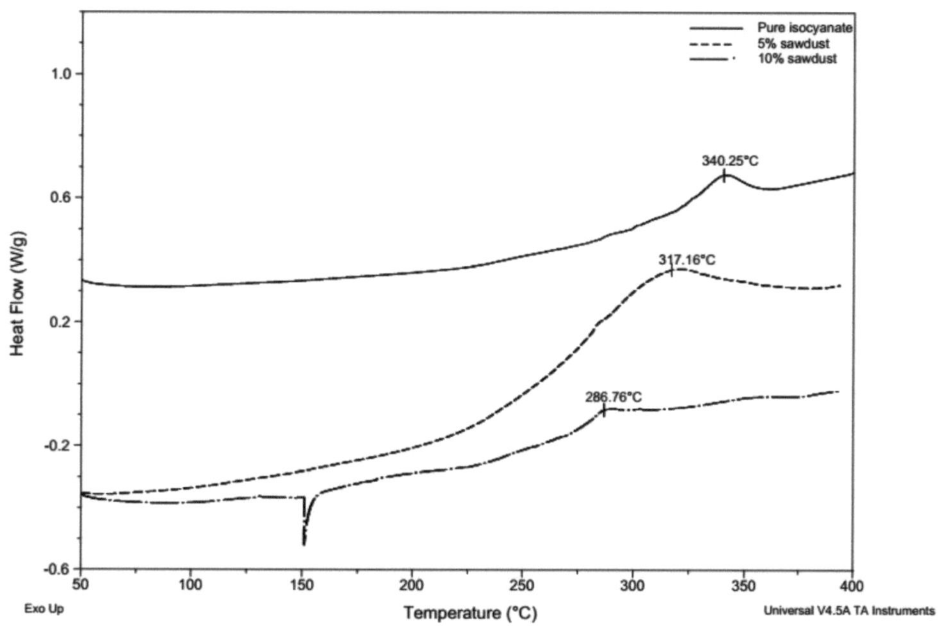

Figure 9. DSC thermograms of pure isocyanate and its mixture with sawdust at different addition levels [43].

$$R{-}N{=}C{=}O + O{=}C{=}N{-}R \longrightarrow R{-}N{=}C{=}N{-}R + CO_2$$

Figure 10. Self-polymerization of isocyanate (–NCO) to form polymeric carbodiimide upon elimination of carbon dioxide [45].

Sample	Mass (mg)	β (°C/min)	T_o (°C)	T_p (°C)	ΔH (J/g)
Pure isocyanate	7.0	10	325.86	340.25	7.31
5% sawdust	6.3	10	300.76	317.16	5.44
10% sawdust	8.5	10	280.98	286.76	1.69

Source: Nuryawan and Alamsyah [43].

Table 4. Results of DSC analysis of isocyanate and its mixture with sawdust at different addition levels.

The result showed that polymeric isocyanate adhesive had a peak temperature (T_p) at 340.25°C. That was probably the self-polymerization temperature of polymeric isocyanate. It is known that self-polymerization of polymeric isocyanate occurs at temperatures above 300°C [44]. The reaction forms polymeric carbodiimide upon elimination of carbon dioxide. It is an exothermic reaction and can accumulate heat in the products. The reaction equation is shown in **Figure 10** [45].

Incorporation of sawdust into polymeric isocyanate adhesive decreased the T_p. As can be seen in **Table 4**, 5% addition of sawdust into polymeric isocyanate adhesive decreases the T_p to 317.16°C. Further addition, up to 10%, of sawdust decreased the T_p drastically to 286.76°C.

Polymeric isocyanates are known to react readily with hydrogen atom (H) in water and alcohols. It seems an addition of sawdust to polymeric isocyanate provides more active H in the system and accelerates the cure of polymeric isocyanate adhesive. The active H obviously originates from the water in sawdust. The moisture content of sawdust used was 6.2%. An increase of sawdust content resulted in greater water content in the adhesive mixture, which eventually reacted with polymeric isocyanate adhesive and decreased the T_p. Moreover, it is known that the polymeric isocyanate adhesive also reacts with hydroxyl groups (–OH) from wood. This type of reaction produces urea or urethane bond as can be seen in **Figure 11** [46].

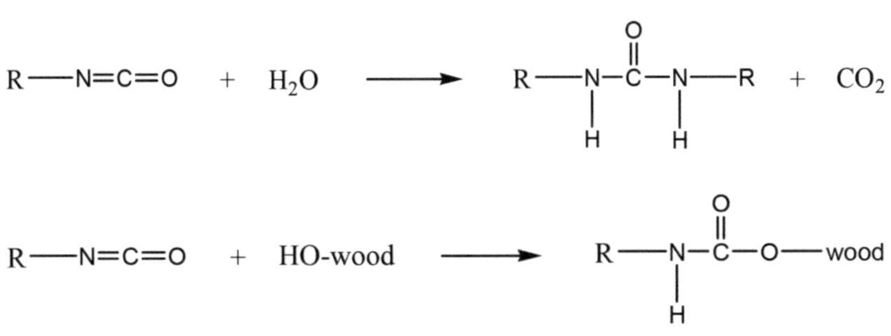

Figure 11. Reaction of isocyanate (–NCO) with water and wood containing water resulted in strong bonds and release of carbon dioxide [43].

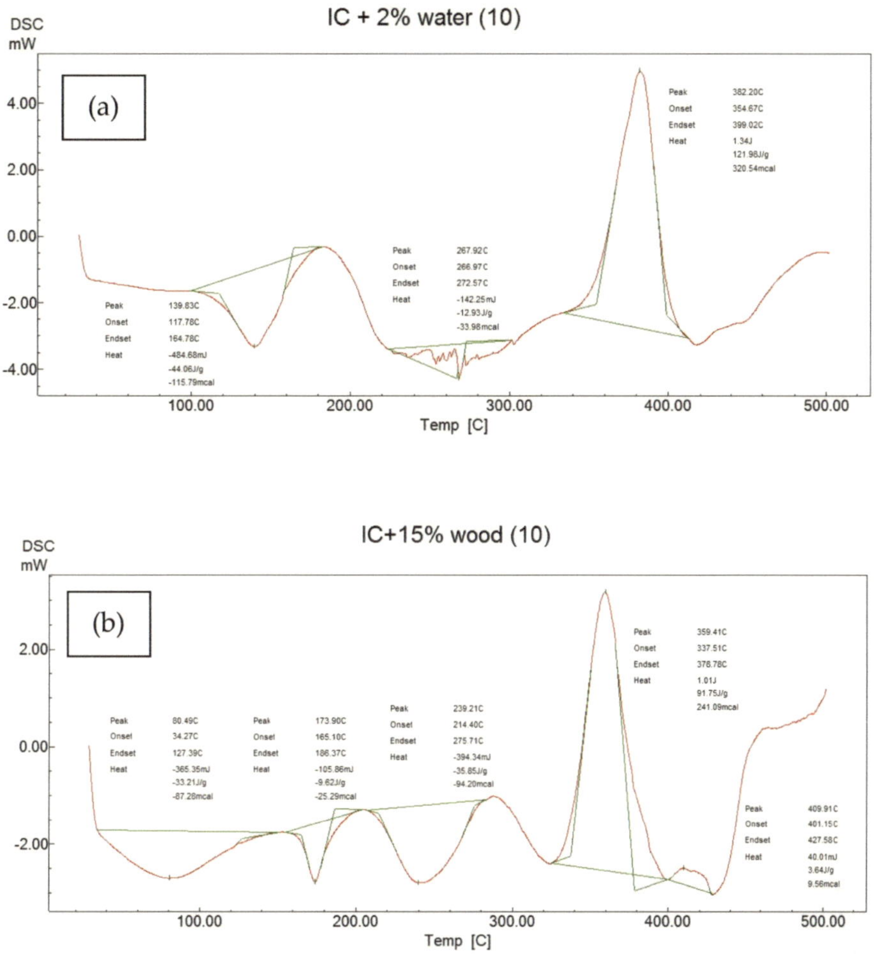

Figure 12. DSC thermogram of isocyanate films; (a) film cured with 2% water loading and (b) film cured with 15% wood particle loading [43].

When isocyanate is used as an adhesive in the particleboard system, the reactive –NCO group reacted with water (because wood is a hygroscopic material and contains free water and bound water) and also with the –OH group from wood (same as the reaction described earlier). Therefore, to study the curing behavior of the isocyanate curing using either water or wood in the form of sawdust, we scan film isocyanate curing using either water or wood under DSC as shown in **Figure 12**.

According to the peaks shown in the thermogram of the DSC scan (**Figure 12)**, there were other components within the isocyanate that have different molecular weight (**Figure 12a**); also wood component reacted with the isocyanate (**Figure 12b**).

Therefore, advance analysis such as gas chromatography-mass spectrometry (GCMS) is needed to clarify this phenomenon. Example of GCMS analysis of isocyanate H3M®

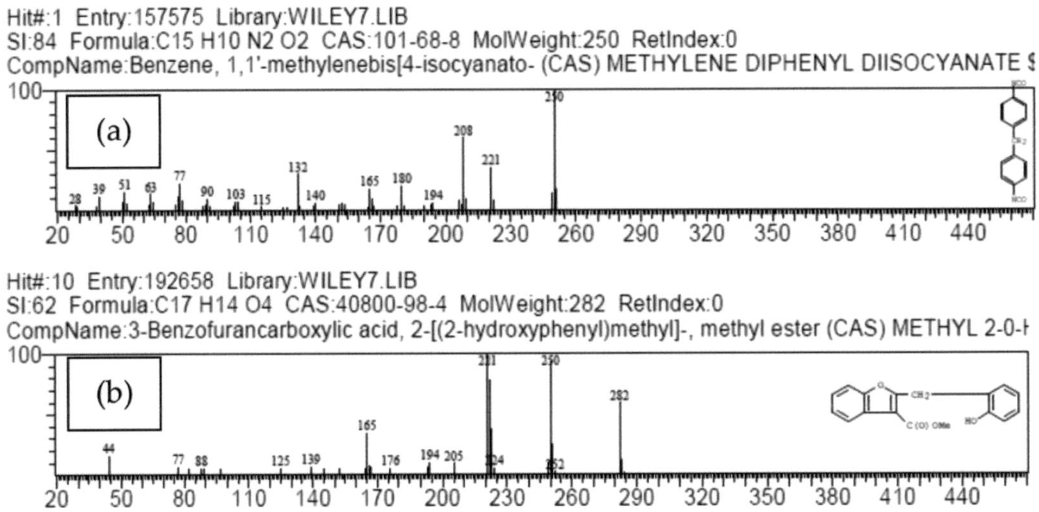

Figure 13. Analysis GCMS of isocyanate H3M® resulted in difference in species and molecular weight such as (a) MW = 250 and (b) MW = 282 [43].

resulted in not only difference in the molecular weight but also chemical species as shown in **Figure 13**.

7. Summary

According to the best of our knowledge, two manufactures in Indonesia, namely PT.Koyolem Indonesia and PT.Polychemie Asia Pacific, have been producing WBPI adhesive consisting of aqueous PVA and isocyanate cross-linker. For research needs, WBPI is used as an adhesive for glulam and LVL production while isocyanate cross-linker (alone) with the trade name H7® and H3M® has been used for bonding in OSB, particleboard, and fiberboard. For optimizing the properties of wood products bonded by isocyanate, the thermal properties have to be investigated such as curing behavior; peak temperature resulted in curing by either water or hygroscopic wood, or high heating for de-blocking isocyanate blocking. Another analysis such as GCMS is important for clarifying the molecular weight of species within the adhesive/glueline (film) system.

Acknowledgements

This work was supported by the Ministry of Technology and Higher Education Republic of Indonesia through the postdoctoral research scheme in the year of 2017 for financial support.

Author details

Arif Nuryawan[1]* and Eka Mulya Alamsyah[2]

*Address all correspondence to: arif5@usu.ac.id

1 Department of Forest Products Technology, Faculty of Forestry, University of Sumatera Utara, Medan, Indonesia

2 School of Life Sciences and Technology, Institut Teknologi Bandung, Bandung, Indonesia

References

[1] Frihart CR. Chapter 9 Wood adhesion and adhesives. In: Rowell RM, editor. Handbook of Wood Chemistry and Wood Composites. Boca Raton, FL: CRC Press; 2005

[2] Matting A, Brockmann W. Recent development in the adhesives field. Angewandte Chemie International Edition. 1968;**7**(8):598-605

[3] Frihart CR, Hunt CG. Chapter 10 Adhesive with wood materials. In: Bond Formation and Performance. General Technical Report FPL-GTR-190. Wood Handbook. USDA, Madison, WI, USA; 2010. pp.10.1-10.24

[4] Ferdosian F, Pan Z, Gao G, Zhao B. Bio-based adhesives and evaluation for wood composites application. Polymer. 2017;**9**(70):1-29

[5] Rowel RM, Ellis WD. Bonding of isocyanates to wood. In: Edward KS, Gum WF, ACS Symposium Series, editors. Urethane Chemistry and Applications. Washington: American Chemical Society; 1981

[6] John WE. Isocyanates as wood binders—A review. Journal of Adhesion. 1982;**15**:59-67

[7] Grøstad K, Pedersen A. Emulsion polymer isocyanates as wood adhesive: A review. Journal of Adhesion Science and Technology. 2010;**24**:1357-1381

[8] Papadopoulos AN, Hill CAS, Traboulay E, Hague JRB. Isocyanate resins for particleboard: PMDI vs EMDI. Holz als Roh-und Werkstoff. 2002;**60**(2):81-83

[9] Frazier CE. Chapter 33 Isocyanate wood binder. In: Handbook of adhesive technology, 2nd ed, revised and expanded. New York: Taylor & Francis Group, LLC. Marcel Dekker, Inc; 2003

[10] Kenji U, Etsuya Y. Production method for particleboard and fiberboard. European Patent Application EP 2 666 609 A1; 2013

[11] Herawati E, Massijaya MY. Balok Laminasi. Cooperation between Faculty of Forestry Bogor Agricultural University and Faculty of Agriculture University of Sumatera Utara. 2006. ISBN 979-25-5472-6 (*In* Bahasa Indonesia)

[12] Koyo Sangyo Co.Ltd. Corporate Profile. 2005. http://www.koyoweb.com/en/company/profile.html [retrieved on January 9, 2018]

[13] Polychemie Asia Pacific Permai. Our History. 2011. http://www.polychemie.co.id/?page_id=14 [retrieved on January 9, 2018]

[14] Hu H, Liu H, Zhao J, Li J. Investigation of the adhesion performance of aqueous polymer latex modified by polymeric methylene diisocyanate. Journal of Adhesion. 2006;**82**:93-114

[15] Herawati E, Nugroho N. Performance of glued-laminated beams made from small diameter fast-growing tree species. Journal of Biological Sciences. 2010;**10**(1):37-42

[16] Alamsyah EM, Nan LC, Yamada M, Taki K, Yoshida H. Bondability of tropical fast-growing tree species I: Indonesian wood species. Journal of Wood Science. 2007;**53**:40-46

[17] PTLemindo Abadi Jaya. Technical Assistance (*In* Bahasa Indonesia); 2003

[18] Vick CB. Adhesive bonding of wood materials. In: Wood Handbook: Wood as an Engineering Material. Madison: USDA Forest Service, Forest Products Laboratory; 1999. pp. 9.1-9.24. General Technical Report FPL: GTR-113 http://www.treesearch.fs.fed.us/pubs/7139

[19] Alamsyah EM. Shear Strength and Wood Failure Percentages of Some Tropical Fast-Growing Wood Species Bonded with Isocyanate, API KR-7800. Unpublished work. Indonesia: Institut Teknologi Bandung; 2017

[20] Conner AH. Wood: Adhesive. In: Encyclopedia of Materials: Science and Technology. New York: Elsevier Science Ltd. 2001. pp.9583-9599

[21] Lepene BS, Long TE, Meyer A, Kranbuehl. Moisture-curing kinetics of isocyanate prepolymer adhesives. Journal of Adhesion. 2002;**78**:297-312

[22] He G, Yan N. Effect of moisture content on curing kinetics of pMDI resin and wood mixtures. International Journal of Adhesion & Adhesives. 2005;**25**:450-455

[23] Glasser WG, Saraf VP, Newman WH. Hydroxy propylated lignin-isocyanate combinations as bonding agents for wood and cellulosic fiber. Journal of Adhesion. 1982;**14**:233-255

[24] BASF. Polyurethane MDI Handbook. Geismar, LA. USA: BASF Corporation; 2000

[25] Zhao LF, Liu Y, Xu ZD, Zhang YZ, Zhao F, Zhang SB. State of research and trends in development of wood adhesives. Forestry Studies in China. 2011;**13**(4):321-326

[26] Nuryawan A. Physical and mechanical properties of oriented strand board made from small diameter akasia (*Acacia mangium*Willd.), ekaliptus (*Eucalyptus* sp.) and gmelina (*Gmelinaarborea*Roxb.). [Thesis].Bogor: School of Postgraduate; 2007

[27] Nuryawan A, Tambunan DH, Hakim L, Alamsyah EM. Physical and mechanical properties of fibreboard made of acacia fibers and isocyanate. Poster presented in IUFRO INAFOR Joint International Conference. July 24-27. Yogyakarta. Conference Programme Book pp.193 IUFRO-INAFOR Joint International Conference 2017

[28] Nuryawan A, Tambunan DH, Hakim L. Isocyanate and their application for adhesive of acacia fibreboard. Paper presented in National Seminar of Indonesian Wood Researchers Society XIX. October 20, Ambon, Maluku, Indonesia: University of Pattimura; 2016

[29] Polyoshika. To Produce Better Laminated Wood. Jakarta: PT. Polyoshika; 2000

[30] Nuryawan A, Risnasari I, Sinaga PS. Physical and mechanical properties of particleboard made of logging residue. Jurnal Ilmu dan Teknologi Hasil Hutan. 2009;**2**(2):57-63 (In Bahasa Indonesia)

[31] Nuryawan A, Azhar I, Situmorang R. Physical and mechanical properties of particleboard made of wood industry waste. Paper presented in XXIII IUFRO World Congress, August 23-28, Seoul, South Korea; 2010

[32] Febrianto F, Hidayat W, Samosir TP, Lin HC, Song HD. Effect of strand combination on dimensional stability and mechanical properties of oriented strand board made from tropical fast growing tree species. Journal of Biological Sciences. 2010;**10**(3):267-272

[33] Febrianto F, Sahroni HW, Bakar ES, Kwon GJ, Kwon JH, Hong SI, Kim NH. Properties of oriented strand board made from betung bamboo (*Dendrocalamus asper* (Schultes.f) Backer ex Heyne). Wood Science Technology. 2012;**46**:53-62

[34] Iswanto AH, Azhar I, Supriyanto SA. Effect of resin type, pressing temperature and time on particleboard properties made from sorghum bagasse. Agriculture, Forestry and Fisheries. 2014;**3**(2):62-66

[35] Six C, Richter F. Isocyanates, organic. In: Elvers B, editors. Ullmann's Enylcopedia of Industrial Chemistry. Weinheim: Wiley-VCH Verlag GmbH & Co.KGaA; 2012. DOI: 10.1002/14356007.a14_611

[36] [WHO] World Health Organization. Diphenylmethane diisocyanate. Concise International Chemical Assessment Document 27. United Nations Environment Programme, the International Labour Organization, and the World Health Organization. Geneva; 2000

[37] National Center for Biotechnology Information. PubChem Compound Database; CID=7570. 2017. https://pubchem.ncbi.nlm.nih.gov/compound/7570 [accessed January 9, 2018].

[38] Tan R. The use of pMDI resin in MDF manufacture. Faculty of Forestry. Canada: University of British Colombia; 2012

[39] Situmorang R. Physical and mechanical properties of particleboard made of wood industry waste. Unpublished thesis. Faculty of Agriculture. University of Sumatera Utara. Medan, North Sumatera, Indonesia; 2011

[40] Lubis MAR, Park BD, Lee SM. Modification of urea-formaldehyde resin adhesives with blocked isocyanates using sodium bisulfite. International Journal of Adhesion and Adhesives. 2017;**73**:118-124

[41] Wicks DA, Wicks ZW. Blocked isocyanates III-Part B: Uses and applications of blocked isocyanates. Progress Organic Coating. 2001;**41**:1-83

[42] Lou C, Di M. Study on cross-linking agent of a novel one-component API adhesive. Journal Adhesion Science and Technology. 2014;**27**:2340-2351

[43] Nuryawan A, Alamsyah EM. Advantages and disadvantages of using isocyanate adhesive; an analytical study on the thermal properties. Paper presented in the 9th International Symposium of Indonesian Wood Research Society; 26-29 September. Denpasar; 2017

[44] Saunders HJ, Frisch KC. Polyurethanes: Chemistry and T echnology, Part 1 High Polymers. New York: Interscience; 1962

[45] Sato Y, Okada K, Akiyoshi M, Murayama S, Matsunaga T. Diphenyl methane diisocyanate self-polymerization: Thermal hazard evaluation and proof of runaway reaction in gram scale. Journal of Loss Prevention in the Process Industries. 2011;**24**(5):558-562

[46] Weaver FW, Owen NL. Isocyanate-wood adhesive bond. Applied Spectroscopy. 1995;**49**(2): 171-176

Section 3

Adhesive Bonding in Medical Applications

Chapter 6

Silicone Adhesives in Medical Applications

Gerald K. Schalau II, Alexis Bobenrieth,
Robert O. Huber, Linda S. Nartker and
Xavier Thomas

Additional information is available at the end of the chapter

http://dx.doi.org/10.5772/intechopen.71817

Abstract

This chapter will review silicone based adhesive technologies, applications and characterization, emphasizing those self-adhesive materials often used in skin contact applications including transdermal drug delivery and wound care device attachment. The silicone pressure sensitive adhesives used in transdermal applications today are thermoplastic and based on silicone polymer and silicate resin chemistries. Previous research has suggested that some drugs readily diffuse through silicone adhesives, prompting their use in transdermal patches. A recently developed silicone acrylate hybrid adhesive technology combines polyacrylate and silicone molecular structures to form a stable, semi-interpenetrated network. This technology provides ease in formulating transdermal drug delivery systems through improved physical stability over simple blends of acrylate and silicone adhesives. The ability of some silicone adhesives to affix bandages without disrupting the wound bed upon removal has led to the wide acceptance of a third type of silicone adhesive technology that unlike the aforementioned thermoplastic materials is thermoset. This adhesive form is based on a platinum catalyzed, cross-linking reaction between vinyl functional and silicon-hydride functional silicone polymers. The various silicone adhesive types have been characterized *via* classical measurements of physical performances. Rheological techniques elucidated herein provide further understanding of the structure-property relationships previously unavailable using classical characterization approaches.

Keywords: silicone, pressure sensitive adhesive, soft skin adhesive, transdermal, wound care, silicone acrylate, polydimethylsiloxane, semi-interpenetrating network

1. Introduction

The term "silicone" is not always used consistently, and should only be used to refer to polymeric materials, avoiding the relatively common confusion with the metallic element silicon (Si). Silicones are synthetic polymers containing Si—O—Si bonds and are used in many industries for their water repellency, ability to wet-out surfaces, high permeability to gases, stability in extreme temperatures, and resistance to thermal, radiation and chemical degradation. The variety of physical forms and physiochemical properties that silicones can display has led to their adoption in a diverse array of healthcare applications including medical devices and as active pharmaceutical ingredients (API) and excipients in medicines for over 60 years [1]. One class of silicone materials that has generated continued interest and research is silicone adhesives, specifically those self-adhering materials that do not require any activation immediately prior to use. Silicone adhesives are used as excipients in transdermal patches, and as skin contact adhesives in prosthetic and wound care device attachment. Recent investigations support the use of silicone based pressure sensitive adhesives for their skin-friendliness, but also to enhance the efficacy of the drug in transdermal drug delivery patch products. Recent silicone technologies like silicone based hybrid pressure sensitive adhesives promise potential performance advantages and improved drug delivery efficacy in transdermal drug delivery systems. Other silicone adhesive types are well known for their atraumatic removal from skin - an ability to remove cleanly from compromised skin without negatively impacting the wound healing process.

This chapter will review silicone based adhesive technologies, applications and characterization, emphasizing those self-adhesive materials often used in skin contact applications. One type of silicone adhesive that is well established in the medical device industry but outside the scope of this work are room temperature vulcanizing (RTV) sealants. While these sealants are an interesting and useful class of materials, they will not be a focus of this chapter. Unlike the self-adhering adhesives discussed in this chapter, once fully crosslinked, the RTV sealants are non-tacky and rubbery and designed to form a permanent bond between substrates. These materials have a similar chemistry to silicone caulks commonly in the construction industry, and have found utility adhering materials to silicone elastomers, bonding parts of medical devices together, and acting as encapsulants and sealants in a variety of medical devices, including pacemakers [2].

2. Silicone chemistry

While the term "silicone" persists in common vernacular, "polyorganosiloxane" is a more appropriate term, and has found acceptance in most scientific literature. Polyorganosiloxanes are organosilicon polymers, the most common of which are the trimethylsiloxy-terminated polydimethylsiloxanes (**Figure 1**) [3].

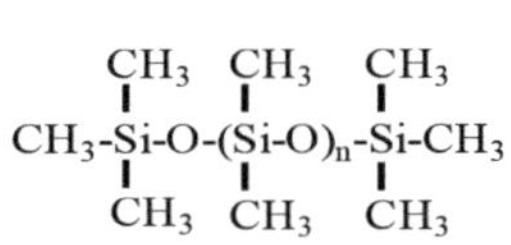

Figure 1. Chemical structure of typical polydimethylsiloxanes.

The silicon in polyorganosiloxanes can be combined with one, two or three organic groups, commonly $-CH_3$, $-CH=CH_2$ or $-H$, with the remaining valence(s) satisfied with oxygen [4]. Branched silicone structures are made possible by substitution of dimethyl siloxane units (i.e., $(CH_3)_2SiO_{2/2}$) with those that contain additional Si—O connections (e.g., $CH_3SiO_{3/2}$ or $SiO_{4/2}$) [4]. It is through the fact that different siloxane units can be combined with one another in the same molecule that the great variety of silicone compounds arises [3].

Silicones exhibit an inorganic backbone chain $(Si\text{—}O)_n$ and organic, (typically methyl) side groups [5]. It is this unusual combination and the resulting physiochemical properties that are responsible for many characteristics of the silicone adhesives. The silicon to oxygen bonds of the backbone are longer and more open than carbon to oxygen bonds permitting the characteristic flexibility of the siloxane chain. By way of comparison, the rotational energy around a $-CH_2\text{—}CH_2$ bond is over four times greater than that of a typical $(CH_3)_2Si\text{—}O$ bond. This flexibility is responsible for the characteristic low surface tension observed in silicones which allows them to quickly "wet out" onto surfaces including skin [5].

In addition to increased flexibility, the silicon-oxygen bonds are also stronger than carbon-carbon bonds. The bond energy of a Si—O bond along the backbone of a silicone polymer is 452 kJ/mol while the typical C—C bond of the backbone of an organic polymer is only about 348 kJ/mol [5]. The inherently strong backbone of silicone polymers can help explain the acknowledged chemical stability silicone polymers possess toward a variety of degradation routes including moisture, UV, and a wide range of temperatures. This is equally important at very low and very high temperatures, where some types of silicones maintain their characteristic physical properties and utility from −100°C up to 260°C [6].

Silicones in general, are hydrophobic, (i.e., having little or no affinity for water), so one may anticipate silicones to be extremely lipophilic, given the common perspective equating hydrophobicity with lipophilicity (i.e., having a strong affinity with lipids). However, in the case of silicones, only relatively small silicone polymers are lipophilic. Polydimethylsiloxane (PDMS) polymers in excess of six to eight $(CH_3)_2SiO$ units have little affinity with lipids while larger polymers are essentially lipophobic. These hydrophobic and lipophobic properties impact the ability to solubilize drugs, oils, botanicals and other traditional active ingredients into a silicone matrix [4]. The relatively poor miscibility of silicones with many compounds may be a key to the noted release efficiency of those same compounds from silicones.

3. Silicone pressure sensitive adhesive: description and applications

Silicone pressure sensitive adhesives (PSA) are comprised of high molecular weight silanol-functional silicone polymers and silanol functional MQ siloxane resins. While a simple mixture of silicone polymer and resin can yield an adhesive with adequate peel adhesion and tack properties, sufficient cohesive strength is lacking. The silicone pressure sensitive adhesives most often used in medical applications are the product of a silanol condensation reaction between the polymer and resin components yielding a network structure, commonly referred

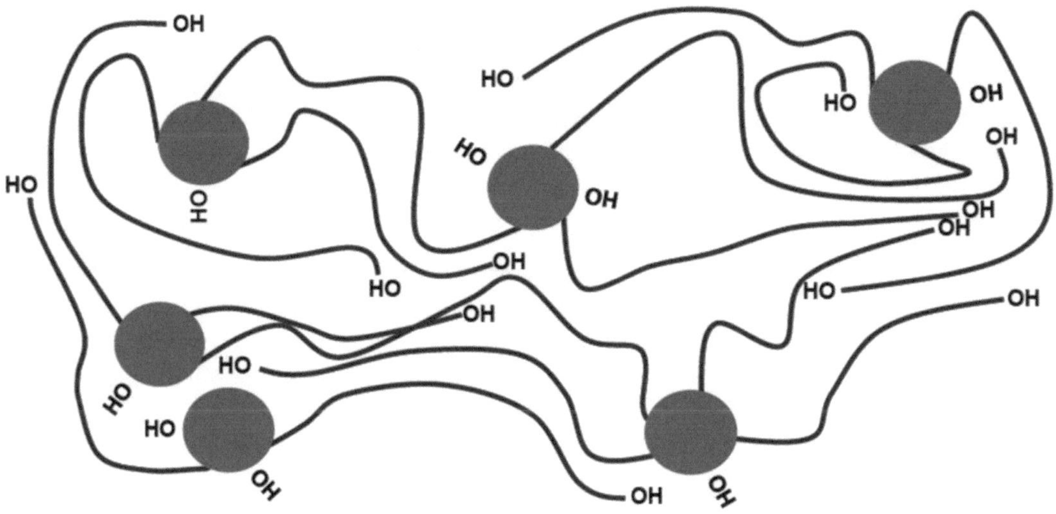

Figure 2. Schematic of the standard silicone PSA.

to as standard pressure sensitive adhesives (**Figure 2**). These materials have suitable cohesive strength for medical device and transdermal drug delivery system applications, and upon removal from the skin the adhesive layer is removed intact. These adhesives are typically supplied in a volatile solvent which is removed during the coating process.

Silicone PSA have a long history of use in transdermal drug delivery systems but may also be used to attach prostheses and wound care devices. One recent innovative example of the utilization of silicone PSA in medical device attachment is the Embrace® MINIMIZE Silicone Scar Aid which consists of a silicone PSA coated onto silicone elastomer (rubber) sheeting. A unique applicator allows the dressing to be applied to relieve tension on healing skin to minimize scar formation [7, 8].

Another application where silicone PSA have found wide acceptance is in the field of transdermal drug delivery. Second to the active pharmaceutical ingredient (API) or drug, the pressure-sensitive adhesive used in a transdermal drug delivery system can be viewed as the most critical component. Without proper and sustained adhesion to the skin, drug delivery from this dosage form does not occur.

Multiple transdermal drug delivery system (TDDS) designs are reported in the literature and are commercially available including reservoir, matrix, and drug-in-adhesive (DIA) systems; slight variants and combinations of each of these patch designs are also found. The functional requirements of the pressure sensitive adhesives in each patch design can vary with the design. (**Table 1**) [9, 10].

Regardless of the patch design, basic requirements for the adhesive that is in direct contact with the skin include sufficient moisture resistance to stay adhered while perspiring and showering and biocompatibility (i.e., the adhesive must be non-irritating and non-sensitizing

Adhesive functional requirement	Patch construction		
	Matrix with rim adhesive	Reservoir with rate controlling membrane/face adhesive	Drug-in-adhesive
Biocompatibility	+	+	+
Moisture resistance	+	+	+
Acceptable tack	+	+	+
Good adhesion	+	+	+
Good cohesive strength	+	+	+
Adherence to backing layer	+	+	+
Adherence to rate controlling membrane		+	+ (in some cases)
Compatible with drug and excipients	+ (in some cases)	+	+
Permeable to drug and enhancers		+	+
Cold flow resistance	+ (esthetic only)	++	++
Stabilize drug and excipients			+

Table 1. Adhesive functional requirements for common transdermal patch designs.

at a minimum). The adhesive must also have acceptable tack to adhere quickly on contact, good wetting behavior to achieve sufficient adhesion for the duration of wear (typically from 12 h to 7 days) and possess sufficient cohesive strength to enable removal without residual adhesive remaining on the skin. In most transdermal patch designs, the adhesive must also resist cold flow, or creep, the property of an adhesive to deform, especially at ambient temperature prior to use or at skin temperature when in use.

The TDDS design with the most straightforward adhesive requirements is a matrix patch with a rim adhesive layer around the periphery of the patch. In this type of patch design, the adhesive functions are not significantly different from other device attachment applications as the adhesive must simply adhere the patch to the skin for the intended wear period. If the rim adhesive layer comes into contact with the drug loaded matrix layer, the adhesive must also be compatible with the matrix layer components. Resistance to cold flow for a rim adhesive is esthetically pleasing but does not result in unintended drug exposure or impact the drug contact surface area, so is not usually a mandatory function.

Reservoir patch designs are typically characterized by a liquid reservoir compartment with solubilized API separated from the skin contact PSA by a semipermeable membrane. For a reservoir patch design with an adhesive layer across the face of the entire patch, the adhesive must adhere to the membrane and provide adequate adhesion to skin, as well as be compatible with the drug and allow diffusion of the drug and any penetration enhancers to the skin interface. The adhesive properties must be resilient to the drug and enhancer(s) reaching saturation in the adhesive layer.

In a drug-in-adhesive (DIA) patch design, the adhesive plays an even greater role in the overall function of the patch. While this type of patch construction is clearly the easiest to manufacture, many formulation challenges exist, particularly with a monolithic (i.e., single layer) design. In addition to the requirements stated above, the adhesive matrix must also stabilize the API and excipients in either a dissolved or dispersed state, and allow controlled release of the drug and enhancers. Cold flow reduction is even more challenging in monolithic patch designs too, as they commonly require a greater adhesive coat weight than constructs that use face adhesive layers.

It is unlikely that any single, off the shelf, adhesive system can meet the demands for all patch formulations and patch types. Silicone PSA, along with acrylic and polyisobutylene (PIB) PSA, are commonly used in transdermal patch applications. The end-use properties of silicone PSA (tack, adhesion, cohesive strength) can easily be modified or customized by varying the resin-to-polymer ratio, the degree of cross-linking and the residual silanol functionality during preparation. Silicone PSA are soluble in a variety of volatile polar and non-polar hydrocarbon solvents and additional customization may be achieved *via* the solvent in which the silicone PSA is dispersed as well as the concentration of the PSA in solvent. Solvent and concentration may be matched to provide optimal conditions for drug and excipient dissolution for TDDS manufacturing. Hot melt forms of silicone adhesives are also available. The capability to uniquely customize silicone PSA is essential for use in transdermal drug delivery applications and is likely responsible for their use therein. There are instances where more customization is required than can be achieved with standard silicone adhesives. For the silicone chemistry described above and noted in **Figure 2**, it is important to note that exposure to amines and amino-functional drugs and excipients will cause certain silicone PSA to lose tack and their ability to instantly adhere to the skin. Standard silicone PSA can be chemically treated to reduce the silicon-bonded hydroxyl (silanol) content of the adhesive to render the PSA resistant to loss of tack, commonly referred to as amine-compatible silicone adhesives (**Figure 3**) [11].

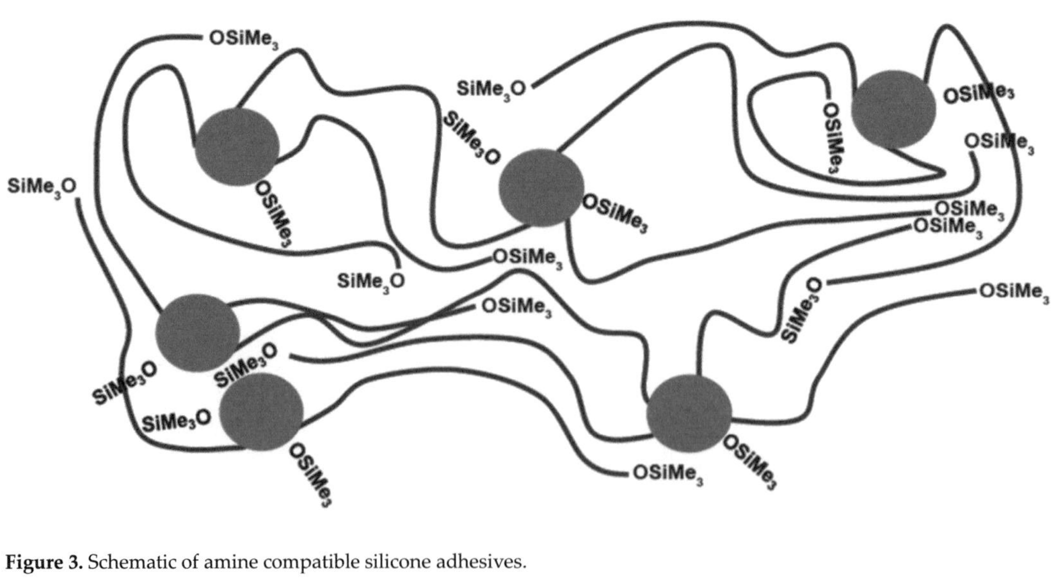

Figure 3. Schematic of amine compatible silicone adhesives.

Silicone PSA is utilized in a variety of marketed TDDS either as the primary adhesive system or in combination with acrylic adhesives. **Table 2** provides a list of commercial TDDS that utilize silicone PSA as a component of the patch construction as of the time of this publication, the respective actives, and other relevant information is also included. The table highlights the evolution of TDDS designs from the first silicone-containing reservoir patch in 1981 to recent approvals of more sophisticated microreservoir and multilayer designs that incorporate different adhesive types to achieve demanding dosage requirements.

In recent years, the nomenclature for silicone PSA listed in the FDA Inactive Ingredient Database (IID) has been standardized to allow patch formulators to more easily identify prior use and maximum potency. Previously, reference to the use of silicone PSA in transdermal patches varied from a description of an adhesive laminate to numeric product codes. The preferred substance name for standard silicone adhesives is now dimethiconol/trimethylsiloxysilicate crosspolymer, and the preferred substance name for amine-compatible silicone adhesives is trimethylsilyl-treated dimethiconol/ trimethylsiloxysilicate crosspolymer. Reference is made to various types of adhesive with the addition of a nominal resin/polymer ratio [12].

Drug	Patch	Marketer	Construction	Silicone PSA components
Nitroglycerin (1981)	Transderm-Nitro®	Novartis	Reservoir	Silicone face adhesive layer
Fentanyl (1990)	Duragesic®	Janssen Pharms	Reservoir	Silicone face adhesive layer
Estradiol (1996)	Vivelle-Dot®	Novartis	Microreservoir monolithic matrix	Silicone matrix adhesive continuous phase with acrylate polymer microreservoirs
Nicotine (1997)	Generic (OTC)	Aveva	Multilayer matrix	Silicone matrix adhesive continuous phase with acrylate face adhesive
Estradiol / Norethindrone Acetate (1998)	CombiPatch®	Noven	Microreservoir monolithic matrix	Silicone matrix adhesive continuous phase with acrylate polymer microreservoirs
Fentanyl (2005)	Generic	Mylan Technologies	Drug-in-adhesive monolitic	Silicone matrix adhesive continuous phase
Fentanyl (2006)	Generic	Lavipharm Labs	Multilayer matrix w/ membrane	Silicone matrix adhesive, continuous phase and face adhesive layer

Drug	Patch	Marketer	Construction	Silicone PSA components
Methylphenidate (2006)	Daytrana®	Noven	Microreservoir monolithic matrix	Silicone matrix adhesive continuous phase with acrylate polymer microreservoirs
Fentanyl (2007)	Generic	Actavis Labs	Reservoir	Silicone face adhesive layer
Fentanyl (2007)	Generic	Mayne Pharma	Reservoir	Silicone face adhesive layer
Rivastigmine (2007)	Excelon® Patch	Novartis	Multilayer matrix	Silicone face adhesive layer
Rotigotine (2007)	Neupro®	UCB	Microreservoir monolithic matrix	Silicone matrix adhesive continuous phase
Capsaicin (2009)	Qutenza®	Acorda	Drug-in-adhesive monolitic	Silicone matrix adhesive continuous phase
Clonidine (2009)	Generic	Aveva	Multilayer matrix w/ membrane	Silicone matrix adhesive continuous phase with acrylate face adhesive
Fentanyl (2011)	Generic	Mallinckrodt Inc	Multilayer matrix w/ membrane	Silicone matrix adhesive, continuous phase and face adhesive layer
Estradiol (2012)	Minivelle®	Noven	Microreservoir monolithic matrix	Silicone matrix adhesive continuous phase with acrylate polymer microreservoirs
Estradiol (2014)	Generic	Mylan Technologies	Microreservoir monolithic matrix	Silicone matrix adhesive continuous phase with acrylate polymer microreservoirs

Table 2. Commercial TDDS patches utilizing silicone PSA.

4. Silicone and acrylate adhesive blends

Silicone and acrylic PSA chemistries as well as combinations of the two are commonly utilized in transdermal drug delivery [13]. The selection of the adhesive is typically drug and TDDS design specific and each adhesive type has its own advantages and disadvantages. Silicone adhesives may be more challenging during patch formulation due to the immiscibility with many drugs and common excipients in the silicone matrix; while acrylic adhesives are often easier to formulate due to the increased solubility of drugs and miscibility of excipients. However, higher drug utilization is often observed from TDDS that employ a silicone PSA over comparable patches that use an acrylic PSA [13, 14]. Yeoh [14] has provided a review of marketed fentanyl patches and has shown patches utilizing silicone adhesives have much greater fentanyl depletion during use and lower residual drug content after their intended use than comparable patches that use an acrylic adhesive. Minimizing the amount of residual drug in the patch at the end of the labeled use period, particularly with opiate drugs, is a focus of a recent FDA Guidance [15].

Combining silicone and acrylic pressure sensitive adhesives to form an immiscible polymer blend can provide benefits for transdermal drug delivery through selective modification of the solubility and/or diffusivity of the drug in the polymer blend matrix [16]. These micro-reservoir systems allow the drug to be solubilized in high concentrations in the discontinuous polyacrylate phase [17] and have been shown to be beneficial in decreasing patch size and required drug loading [18]. This technique has been successfully implemented in several commercial transdermal patches on the market including CombiPatch®, Daytrana® and Minivelle® (Noven Pharmaceuticals) as well as Vivelle Dot® (Novartis Pharmaceuticals) [16] A review of label claims for two patches that provide a 0.5 mg/day dose of estradiol reveals that a 5 cm^2 Vivelle Dot® patch, which employs the Dot Matrix® technology, can deliver 22.4% of the drug, whereas the 12.5 cm^2 Climera® with an acrylic PSA construction only delivers 9.0% of the drug [19]. These immiscible blends do have a major limitation in that they will exhibit macro phase separation in the coating mass if mixing is discontinued which may be exacerbated upon addition of other formulation ingredients such as penetration enhancers [20]. One potential means to prevent macro phase separation of the two immiscible adhesives is to covalently link the two polymer chemistries together, creating a silicone-acrylate hybrid material.

5. Silicone-acrylate hybrid

Hybrid adhesives, in which silicone and acrylic chemistries are combined, have been described following different routes [21, 22]. One approach is the reaction product of a (meth)acrylate-functional silicone PSA and ethylenically unsaturated monomers, [21] whereas a second route toward a hybrid adhesive describes an alkoxysilyl-functional acrylic prepolymer that is further condensed or "bodied" with silicone PSA precursors (i.e., OH-functional silicate resin and OH-terminated PDMS) in the presence of a condensation catalyst [22]. These hybrid adhesives, although produced *via* opposite approaches, likely have the potential for making very similar materials depending on the exact formulation and extent of covalent coupling between the acrylate and silicone phases. As with the simple blends of silicone and acrylic adhesives mentioned above, the hybrid materials result in an immiscible matrix and exhibit a typical domain (droplets of incompatible material) in continuous phase appearance. However, unlike simple blends, the hybrid adhesives are capable of much finer domain sizes and demonstrate superior phase stability during formulation and in a cast film as shown in **Figure 4** [20].

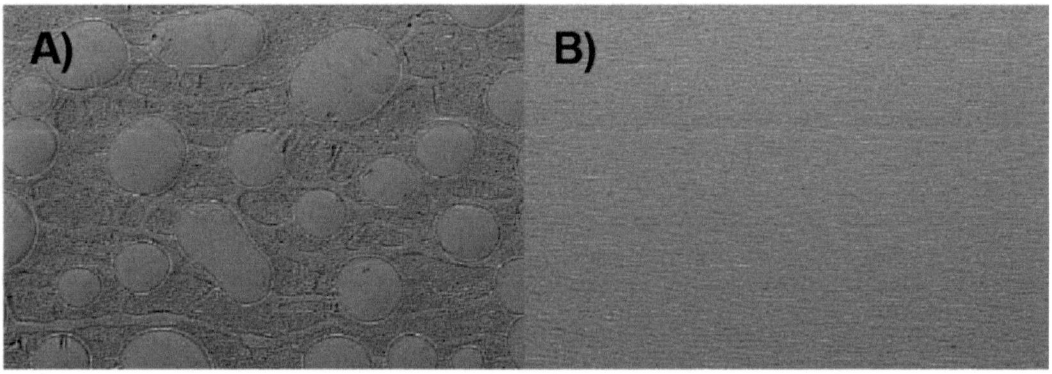

Figure 4. Optical micrograph (100X magnification) of (A) 50:50 blend of silicone PSA and non-functional acrylic PSA and (B) silicone-acrylate hybrid adhesive (50% acrylate) [20].

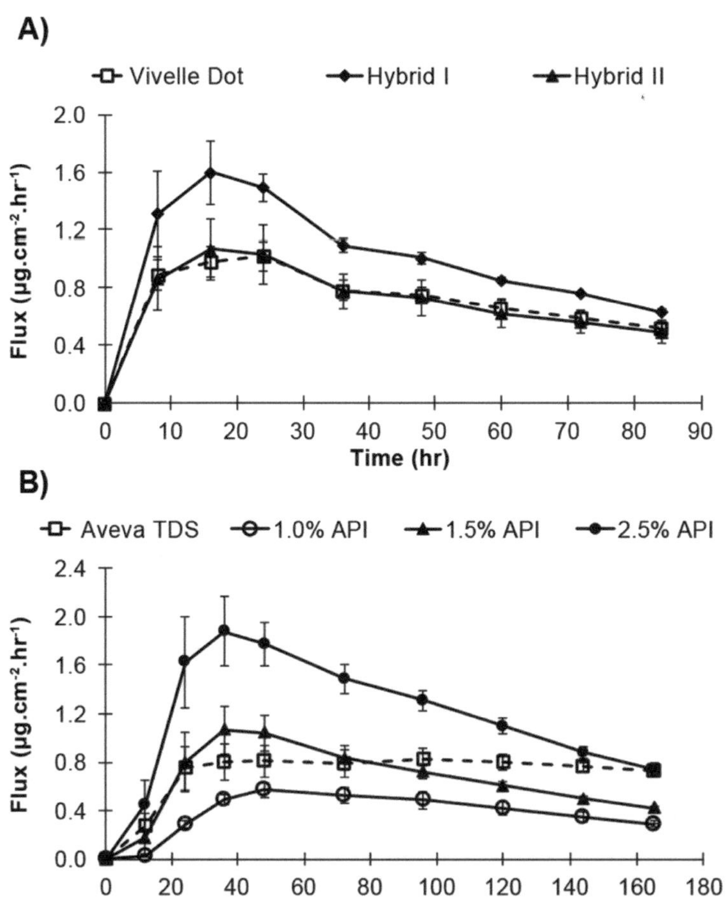

Figure 5. Drug flux from silicone-acrylate hybrid PSA based patches; (A) estradiol 1.5 wt%; (B) clonidine at 1, 1.5 and 2.5 wt%; [23].

Drug delivery using silicone-acrylate hybrid adhesives (SilAc I and SilAc II) differing in the ratio of high and low T_g acrylic monomers has been reported, and delivery of estradiol (**Figure 5A**), clonidine (**Figure 5B**), and ketoprofen was demonstrated across human cadaver epidermis from these matrices. The authors also noted that the use of silicone-acrylate hybrid PSA, singularly or as blends with silicone PSA resulted in a more desirable wet blend compatibility/stability than those obtained with blends [23].

Due to the inherent immiscibility of silicone and acrylate polymers, the hybrid adhesives contain micro-domains which can be observed using transmission electron microscopy (TEM) as presented in **Figure 6**. Further analysis of the phase behavior reveals the ability to selectively control the domain arrangement (i.e., silicone-in-acrylate or acrylate-in-silicone) of these materials by the choice of casting solvent, with the phase having the highest affinity with the casting solvent remaining external, (i.e., heptane casting solvent exhibiting a silicone continuous phase and polyacrylate discontinuous phase (**Figure 6A**) or *vice versa*, (**Figure 6B**)). Phases can also be controlled through changing the volume fraction of silicone or acrylate through blending or addition of specific co-solvents.

The selective control of the phase arrangement provides potential options for tuning both the adhesive properties as well as tailored drug release profiles as illustrated in **Figure 7**.

The impact of casting solvent and silicone content on the material properties has been conducted using a dynamic rheometer (**Figure 8A**). Blends of silicone PSA and silicone-acrylate hybrid PSA (nominally 50% silicone) were prepared in either heptane or ethyl acetate to yield a range of materials. For materials delivered from ethyl acetate, between 76% and 78% silicone, a precipitous change in tan delta is observed followed by incremental decrease as the silicone content rises. Tan delta is a rheological property that approximates the internal friction of a material. When tan delta is greater than one, a material is more viscous than elastic,

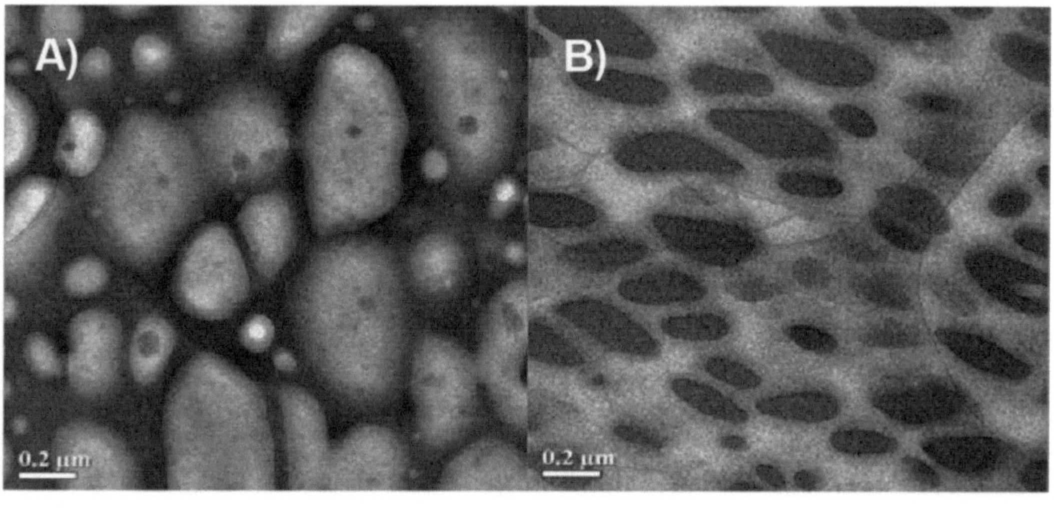

Figure 6. Transmission electron micrograph of silicone-acrylate hybrid adhesive films, silicone phase appears dark due to the electron density (A) cast from heptane and (B) cast from ethyl acetate.

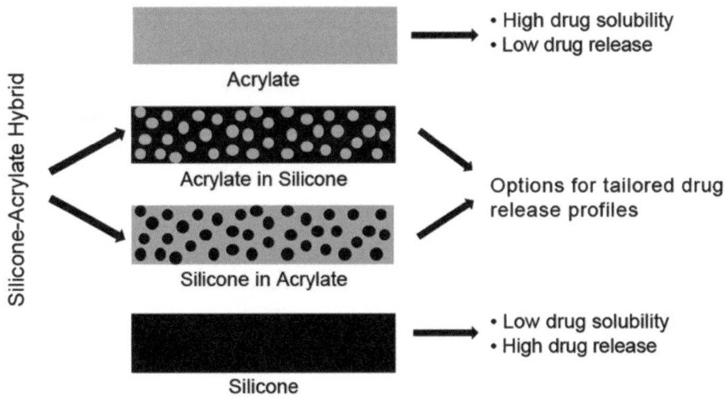

Figure 7. Illustration of silicone-acrylate hybrid adhesive microstructure and potential impact on drug solubility and release.

and when it is less than one it is more elastic than viscous. TEM analysis suggests this is the result of phase inversion when the silicone becomes the external phase. This change is not observed for materials delivered from heptane as the silicone remains the external phase over the entire range. Films containing either 1.0 wt% estradiol (**Figure 8B**), 2.5 wt% ibuprofen (**Figure 8C**), or 2.5 wt% lidocaine (**Figure 8D**) were prepared using blends of hybrid PSA

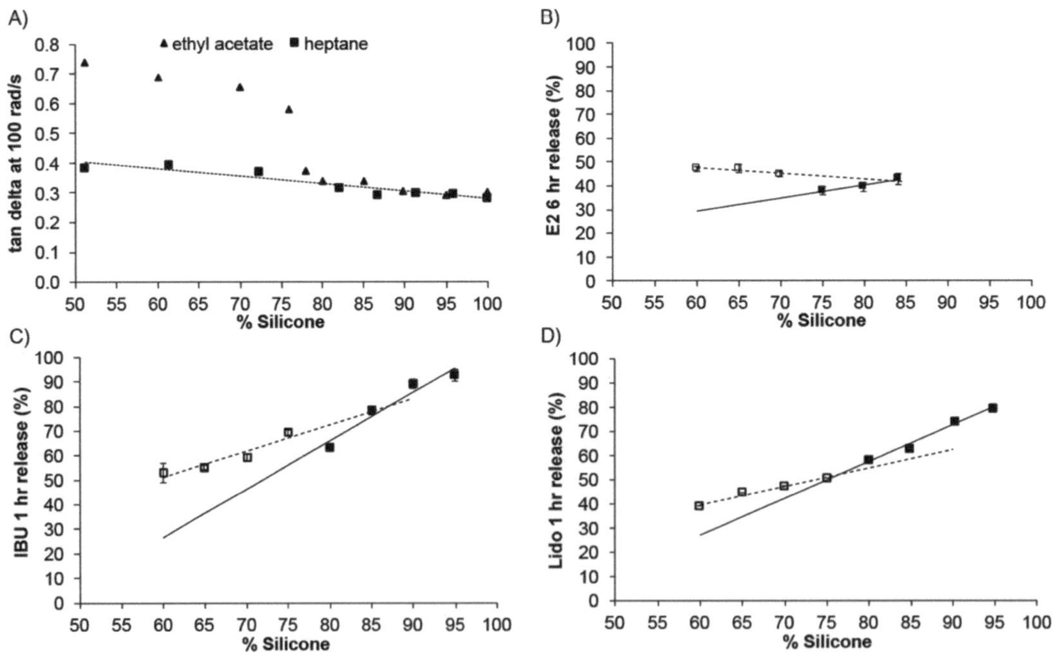

Figure 8. Rheology and drug release as a function of silicone content and dispersion solvent; (A) tan delta of the adhesive matrix; (B) estradiol (E2) 6 h cumulative release; (C) ibuprofen (IBU) 1 h cumulative release; (D) lidocaine (Lido) 1 h cumulative release [24].

with silicone PSA to investigate the impact of phase arrangement on the release behavior. All three API demonstrate a change in drug release characteristics between 75 and 80% silicone content, which is where rheology suggests the phase inversion occurs [24].

6. Silicone pressure sensitive adhesive: strength characterization

The characterization of PSA materials is a critical part of innovation development and production quality control. Historically, tape properties such as peel adhesion, shear and tack have been used to characterize the performance of pressure sensitive adhesives targeted for transdermal applications. However, these tests often have high variability resulting in wide specification limits and poor correlation of test data with adhesive performance in real life applications [25]. Furthermore, tape property tests can be substrate dependent. That is to say, they are influenced by the substrate on which the PSA is coated and also by the substrate on which the adhesive performance is measured. Despite the drawbacks of tape property testing, they are still commonplace and so, warrant some discussion.

Peel tests are well described in the literature and are common to the majority of adhesives. The peel test typically occurs at 90° or 180° and the force to remove the adhesive from a substrate (e.g., stainless steel in many cases) is measured. In the case of silicone PSA, the typical adhesive thickness tested is relatively thin, commonly between two and five mil (approximately 51–127 micron). A distinction between peel adhesion and tack of an adhesive is often made. From an analytical test perspective, the distinction between peel adhesion and tack measurements is the time allowed for the adhesive to bond with the substrate. When measuring tack, the measurement is taken almost instantaneously after the adhesive comes in contact with the test substrate, whereas peel adhesion is measured after the adhesive is left in contact with the substrate for a longer time period. The time between application and testing allows the adhesive to wet out on the surface and the adhesion to build.

Shear testing may have greater relevance to skin contact adhesive applications than the aforementioned peel adhesion and tack tests. Since PSA are condensed materials that have the ability to flow, the extent of cold flow must be characterized to fully understand and anticipate the surface area of adhesive in contact with skin, which can impact the amount of drug delivered from a transdermal patch. Shear tests of fully formulated adhesive matrices may be even more relevant to the performance of the final TDDS. If the skin/adhesive interface changes over time, the transdermal drug diffusion will also change. Typically, a shear test is the measurement of the time for the adhesive to detach from a surface (e.g., stainless steel) under a constant weight.

The advantages of tape property test methodology include ease of set up, reproducibility and a straightforward interpretation of data. However, drawbacks including the considerable influence adhesive coating thickness has on the test, the influence of the substrate on which the adhesive is coated, and the surface on which the test is conducted must also be rationalized. To minimize these influences, there must be accurate control of adhesive thickness and standardization of substrates and test surfaces.

7. Silicone pressure sensitive adhesive: rheology

Although tape property testing may qualitatively predict how quickly a system may bond to a substrate, the extent to which the adhesive resists cold flow, and how much force may be needed to remove it, and perhaps most importantly, the wear performance of the system may not be adequately addressed using classical characterization techniques. In order to better understand and predict the wear performance of transdermal systems, rheology is often used to understand the adhesive bulk viscoelastic behavior. [26] Rheological characterization allows the analyst to overcome the inherent uncertainty linked to peel, tack and shear tests by minimizing the influence of sample preparation and substrate variability on adhesive characterization results. Rheology is a technique to characterize viscoelastic properties of polymers and also predict wear performance of pressure sensitive adhesives. As shown below in **Figure 9**, a typical rheological curve can be correlated to tape properties [27–30].

Data have shown that for viscoelastic materials, such as silicone pressure sensitive adhesives, frequency sweep curves are sensitive to structural differences (e.g., crosslink density) and formulation changes (e.g., resin-to-polymer ratio). This sensitivity provides a means to identify, characterize and predict adhesive wear performance [26].

Storage modulus (G′) is an indicator of how elastic the adhesive is and how much energy is stored during deformation, while the loss modulus (G″) indicates the viscous component of the PSA and how much energy is lost as heat, while complex viscosity (η^*) is an indicator of the adhesive bulk viscosity and can be related to the cold flow [25]. Bonding of a transdermal system occurs at a low deformation rate, and is dependent on the wetting behavior of the adhesive when it comes into contact with skin [26]. Rheologically, the storage modulus, G′, values at low frequency may be used for predicting wetting and creep (cold flow) resistance. Optimum wetting occurs when the adhesive modulus is low. Subsequently, debonding of a transdermal system occurs at high deformation rates [26].

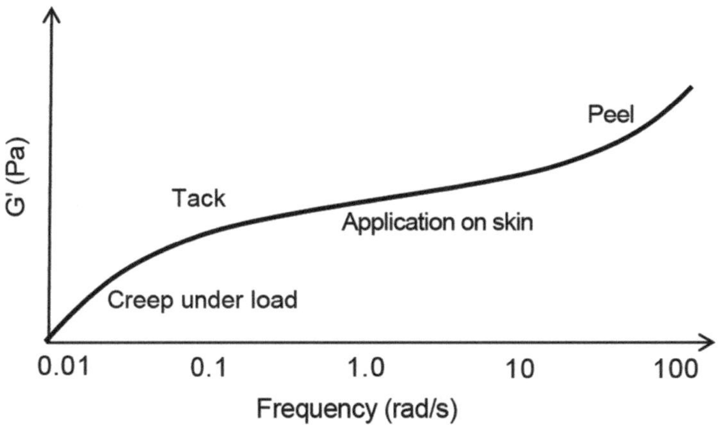

Figure 9. A schematic representation of the link between the rheological profile and the final pressure sensitive (PSA) wear performance [25].

Rheologically, the storage modulus, G′, and loss modulus, G″, at high frequency may be related to the peel adhesion and quick stick (i.e., tack) properties of an adhesive and the subsequent TDDS [31, 32]. For bonding, the viscous contribution should be higher than the elastic contribution to the PSA viscoelastic profile. In rheological terms, this means that at low frequencies, G′ < G″ and the opposite for the debonding step, represented at high frequencies where G′ should be equal to or higher than G″. Based on this interpretation, the rheological traces in **Figure 10** suggest that the increase of resin content should lead to reduced cold flow (i.e., an increase of the complex viscosity with resin content) and an increase of the adhesion strength (i.e., increase of both G′ and G″ with resin content). Dynamic frequency sweeps (0.01–100 rad/s) were conducted on dried adhesive solids using a TA ARES-G2 rheometer. The adhesives with high and medium resin content were tested using 8 mm parallel plates, at 0.35% and 0.5% strain respectively. The adhesive with low resin content was tested using 25 mm plates, at 0.5% strain. All samples were tested at 30°C with a 1.5 mm gap.

In the early 1990s, E.P. Chang developed a theory to interpret rheological data of pressure sensitive adhesives and establish criteria for PSA classification when used in conjunction with the Dahlquist's criteria [33]. This theory is now well known as "Chang viscoelastic window." As depicted in **Figure 11**, a G′ vs. G″ graph, is divided into four quadrants with a central axis. The location of the analyzed PSA within this graph allows a straightforward extrapolation from rheological properties to real-world adhesion performance. For example, the top right hand quadrant corresponds to high modulus and high dissipation. Therefore, materials in this quadrant with characteristically high G′ modulus compensated by the high G″ are anticipated to be adhesive materials with high adhesion but low tack and high shear

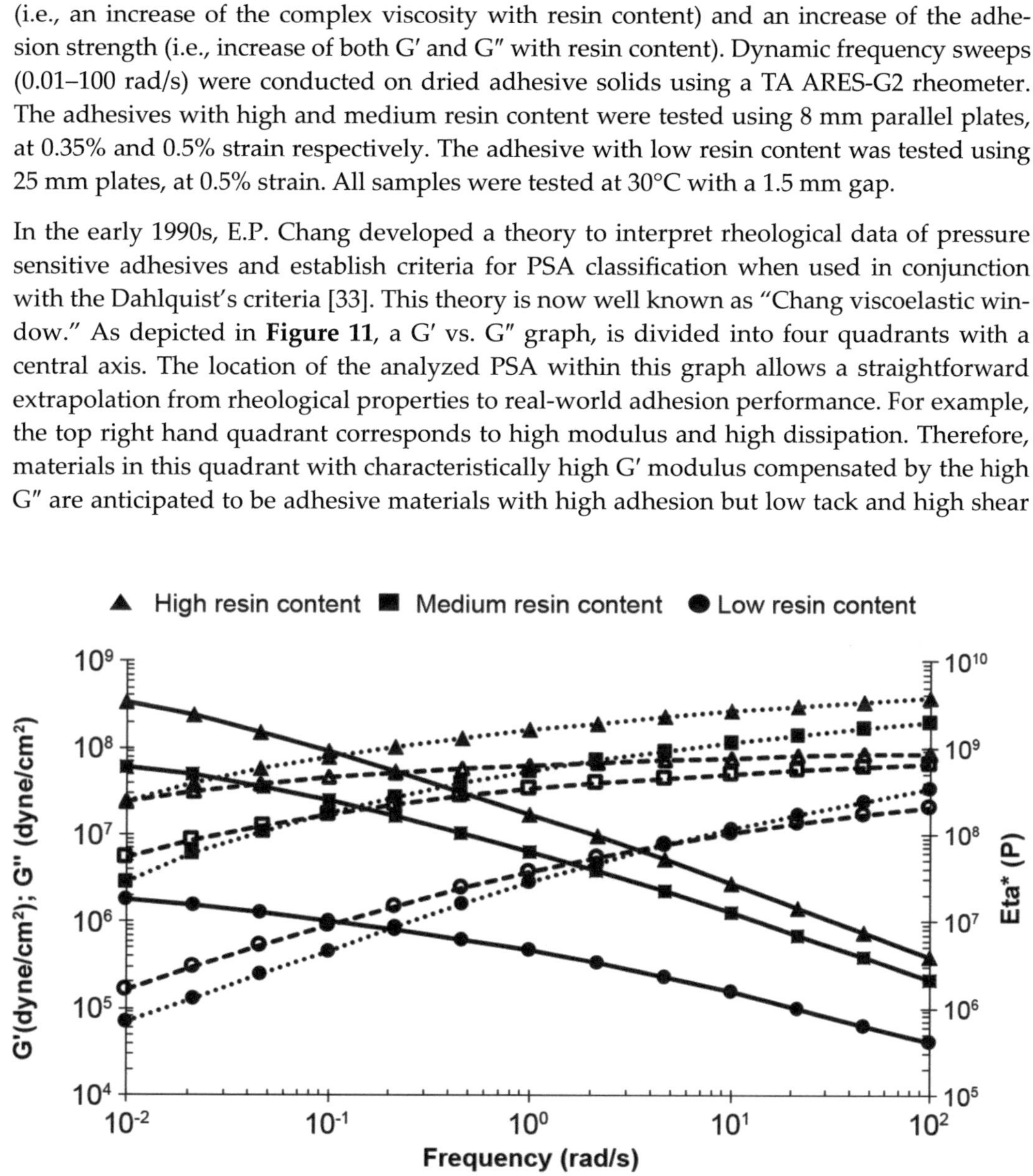

Figure 10. Typical frequency sweeps of silicone PSA at three common resin contents.

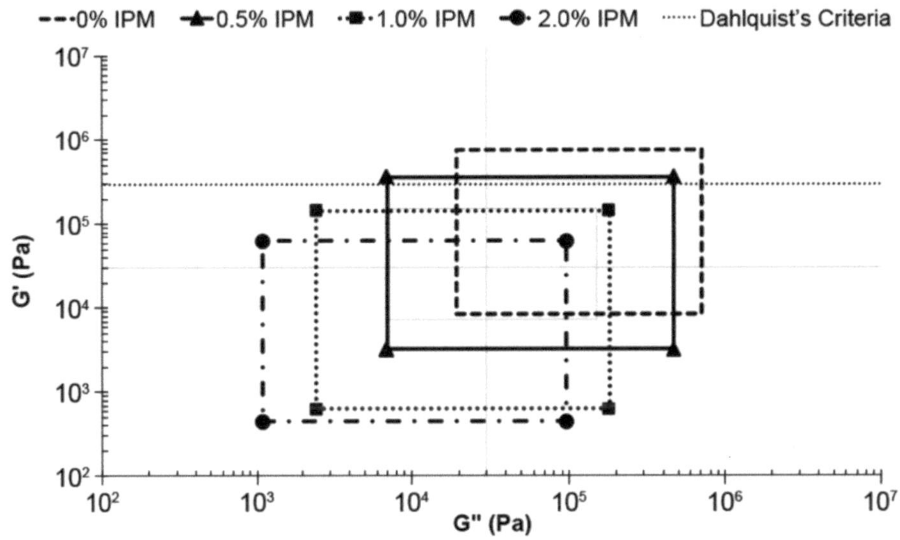

Figure 11. Chang viscoelastic window concept adapted for low resin content silicone pressure sensitive adhesive (PSA) with differing amounts of isopropyl myristate (IPM) [34].

resistance. Conversely, the bottom left quadrant corresponds to low modulus and low dissipation; these materials, are anticipated to exhibit low peel values because of the comparatively low debonding cohesive strength and low dissipation.

Changes in the Chang viscoelastic window, of a typical low resin content silicone PSA can be observed as differing amounts of a commonly used permeation enhancer, isopropyl myristate (IPM), are added (**Figure 11**) [34]. The Chang viscoelastic window of the neat adhesive moves from the upper right quadrant to the lower left quadrant as more IPM is added. The lowermost edge of the window which is linked to bonding of the adhesive is far below Dahlquist's criteria, so the adhesive would be expected to have reasonable tack. There is a significant shift in the position of the upper right corner as IPM content increases which is linked to debonding (peel) efficiency suggesting that an increase of IPM content decreases peel efficiency [34]. Finally, the window size increase indicates a decrease of the PSA shear strength likely due to better solvent compatibility in the PSA. These data coincide with observed changes in adhesive properties as plasticizing agents like IPM are added and support the further use of rheological measurements to characterize changes in wear properties.

8. Silicone soft skin adhesive: description and applications

Silicones have more than 30 year history of safety and efficacy in advanced wound care applications. Much of the success of silicones in wound care is due to an adhesive technology referred to in the literature by many names including soft skin adhesives (SSA), tacky gels, silicone gels and silicone tacky gels among others [35]. The technology was introduced to the wound care market by Dow Corning Corporation in the 1990s and similar materials are offered today by many silicone suppliers under a variety of brand names [36–38]. In a segment that was historically controlled primarily by acrylic adhesives, the tacky gel technology concept was disruptive by securing wound dressings while providing gentle adhesion upon removal. SSAs have become the material of choice in many advanced wound care applications, due to their reliable adhesiveness, while being easier to remove and causing less pain than many other adhesive technologies of the day.

SSAs are based on a polydimethylsiloxane network which supports the critical adhesive attributes required for securing the device in place and removing it without leaving residue or damaging the skin. Unlike silicone PSAs that build their adhesiveness on a viscous phase bodied with a silicate resin, SSAs are based on the silicone elastomer technology modified to deliver the relevant visco-elastic profile. They also differ from analogous silicone elastomers (e.g., liquid silicone rubber (LSR) technology) by the absence of reinforcing silica filler. As a result, they have a similar consistency to gels, but SSAs are not a typical polymeric gel because they are not based on an insoluble polymer network swollen with fluids. The visco-elastic behavior of SSA also differs from silicone PSA, despite their low consistency and a high degree of compressibility, SSAs show resilience and quick recovery under cyclic deformation [35].

The pressure sensitive adhesive property of SSAs are based on the capacity of the elastomer surface to quickly wet the skin and conform to skin irregularities without an additional compression step as required for a silicone PSA [35]. Thanks to the low intensity of the viscous component of the SSA rheological profile, the adhesive does not flow significantly, and very little dissipation of the energy occurs when deformation pressure is applied to the SSA. As a result, SSA debonding happens at low peel force, without skin stripping and painful skin pulling when the adhesive device is removed. Being elastomeric by nature, SSAs have a low viscous component that limits their flow and consequently the ability to pick up materials on or from the surface of the skin [35]. Therefore, unlike silicone PSA, the adhesive surface of SSAs remain relatively clean upon removal from the skin, allowing for removal and easy reapplication of the dressing or device to the skin, making wound dressing repositioning possible.

The elastomeric structure of SSAs is obtained by cross-linking a network of polydimethylsiloxane (PDMS). The reaction is based on an addition reaction (hydrosilylation) between vinyl functional PDMS (polymer) and hydrogen functional siloxanes (cross-linker) as shown in **Figure 12**. The cure reaction is catalyzed by a platinum complex, which can occur at room temperature or be accelerated at elevated temperature (80–145°C), without the formation of reaction by-products [35]. As thermoset materials, SSAs have a low susceptibility to cold flow and plasticizing effects.

The SSA technology has been extensively used in scar treatment and advanced wound management, demonstrating safety and efficacy recognized by wound care professionals [35]. The use of SSA may be recommended when designing medical adhesive devices, tapes, bandages, drapes, and wound dressings and have been noted for the many benefits including high tack for quick bonding to skin, reliable adhesiveness and cohesiveness, gentle adhesion to fragile and compromised skin, no skin stripping and pain-free removal of the device, as well as permeability to moisture and gases (e.g., CO_2, O_2) [35].

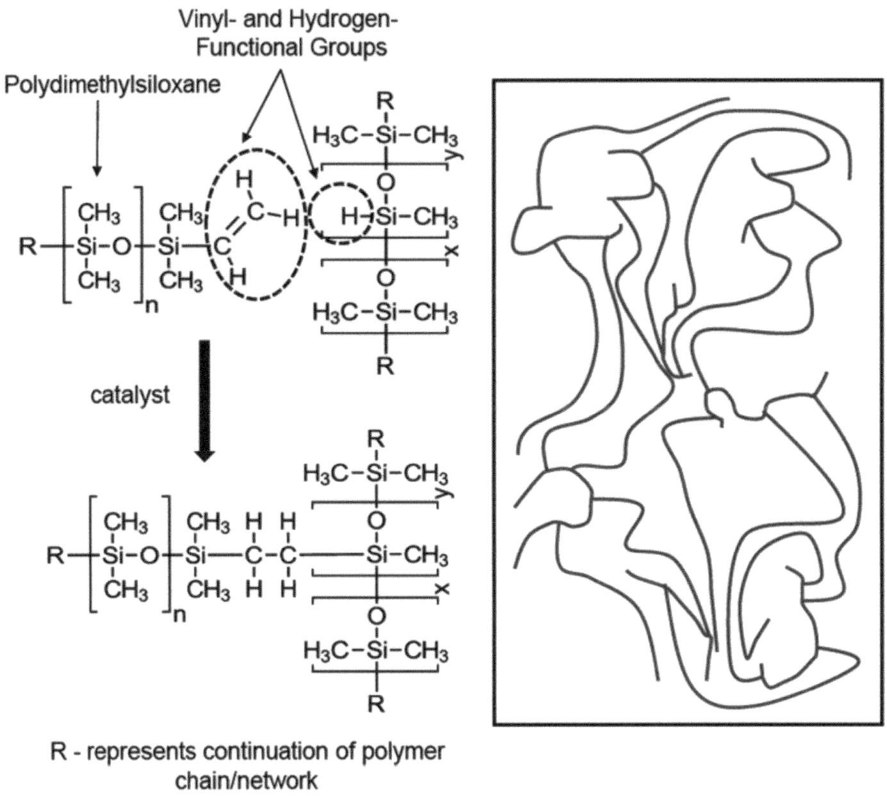

Figure 12. Typical hydrosilylation reaction schematic.

SSAs are supplied as two-part systems with the catalyst in one part and the cross-linker in the other. The materials are characteristically transparent before and after curing into a solid matrix. They are typically processed by mixing the two parts and coating the mixture directly onto the final substrate (i.e., backing film), understanding that this film must be impermeable enough to prevent the uncured liquid SSA from wicking through. The typical coat weight for SSA can vary widely depending on the desired final properties, but often range between 150 and 250 g/m^2. The curing phase is typically completed at elevated temperature adjusted according to the temperature sensitivity of the substrate. After cooling, the adhesive surface is protected by a release liner which is peeled off when the end user applies the adhesive to skin.

Substrate selection is important when designing an adhesive device based on SSA, as the nature of the substrate can significantly impact the coating and cure conditions during the manufacturing phase. The anchorage of the adhesive to the substrate and the cohesion of the adhesive after cure, as well as the ultimate wear behavior of the device when applied to the body can all be impacted by the substrate selection.

The choice of release liner is also a critical factor as it can affect the device stability, making it unusable if this protective film cannot be easily removed from the adhesive prior to use. Traditional silicone release liners that are used ubiquitously with acrylic adhesives cannot be used with SSA as the silicone release liner chemistry is similar enough to SSA that they are highly likely to interact and experience an irreversible lock-up effect upon storage. However, uncoated polyethylene films, especially LDPE (low density polyethylene) grade, can provide an acceptably low and reasonably consistent release force from the SSA [39].

New SSA technology are being developed that can achieve higher adhesion and longer wear times as well as improved drug compatibility to address emerging medical system market trends including wearable devices and topical drug delivery patches [35]. The use of SSA technology to formulate drug delivery matrices enables drug delivery system designs which address the needs for secure and gentle fixation to fragile, sensitive or compromised skin conditions common in dermatology, wound care, pediatrics and gerontology. Several studies were conducted to evaluate the compatibility of various drugs and their release from SSA matrices. A variety of API have been studied including those indicated for pain relief and local anesthesia, antibiotics, and dermatological actives [39]. Wound care products that utilize silicone tacky gels as the skin contact adhesive and are loaded with chlorhexidine gluconate and other antimicrobial agents have also been investigated [40]. This may signal further interest in the utilization of SSA in even more advanced active-loaded therapies in addition to the traditional wound therapies where it has been used historically.

9. Soft skin adhesive: characterization

Many of the analytical techniques used to characterize silicone PSA have been modified to characterize the SSA materials, although shear tests are less emphasized for SSA due to the characteristically low cohesion of the SSA. In addition to adhesive peel measurements, the measurement of the softness of the SSA by penetration test is often performed. Over a broad range, the penetration measurement shows correlation to adhesion performance values within a formulation type and is linked to the adhesive network chemistry; therefore, it is often used as a quality control measurement.

Peel tests are commonly used in the adhesive industry, because for many applications these relatively easy to perform tests fit well with the final application of the adhesive. The substrates upon which most adhesives are tested to evaluate adhesive strength (e.g., stainless steel) often are not predictive of the relative strength SSA will exhibit in practice on skin. Therefore, some users have resorted to using substrates that have a surface energy more similar to that of skin as the test substrate for SSA. The number and diverse composition of substrates including plastic films, paper and even artificial skin materials, make standardization across the industry difficult, and comparison between users problematic. Testing is conducted similarly to that described for PSA, with the SSA typically being cast and cured at a consistent thickness directly onto a film. This substrate may influence the peel adhesion result due to its intrinsic elasticity and also potentially through interactions with the SSA. The gel on the backing substrate is then applied on a test substrate, taking care to apply the adhesive with a constant force. After a designated equilibration time, the adhesive is peeled from the substrate, typically at a 180° angle, and the force required to remove it is measured.

While this method provides relative adhesion strength, allowing comparison of adhesive values, the results may be significantly influenced by the backing substrate, as well as the test substrate used, so the results do not necessarily simulate the application of the adhesive to skin.

10. Soft skin adhesive: rheology

Rheological measurements have been developed and used for decades to characterize silicone PSA and provide more realistic predictions of real-world adhesive performance than classical peel tests are capable of providing. Recently, similar rheological measurements have been applied to characterize the intrinsic properties of the SSA and offer a characterization method more capable of harmonization across the industry. The SSA rheological characterization is performed on free standing gels and is able to characterize the adhesive properties without the influence of backing or test substrates unlike the aforementioned adhesion tests. SSAs may be characterized in dynamic oscillation modes, using strain and frequency sweeps to measure the viscoelastic characteristics (e.g., storage modulus, G′ and loss modulus, G″). Different SSA, which exhibit significant differences with respect to adhesion can also be discriminated using rheological analysis. Identifying the true viscoelastic properties of the adhesives is critical to understand the adhesion performance of such products. Using the data generated from the rheometer, it is possible to correlate viscoelastic properties to adhesion, and to better understand structure-property relationships.

To understand the rheological characteristics of this material one must identify the linear viscoelastic (LVE) zone by submitting the sample to an oscillatory strain sweep analysis. In the LVE zone, the elastic modulus (G′) and the loss modulus (G″) are independent of the shear strain, indicating that within this strain zone, the response of the material does not depend on the strain applied, and there are no modifications of the material structure. In the LVE zone identification test, the strain is the only parameter which varies, all other parameters, (e.g., temperature and oscillation frequency) are fixed. The LVE graph for the SSA exhibits a large linear viscoelastic zone from 0.5 to 30% logarithmic strain, providing some flexibility to set the strain when performing the frequency sweep at a fixed strain is the next step of the measurement process. Knowing the LVE zone of the material allows one to carry out the second phase of the rheological evaluation, the oscillatory frequency sweep test. Previously unreported data is shown to elucidate this concept in **Figure 13**. Samples were prepared by weighing equal amounts (±2%) of the two parts of the SSA and mixed to ensure homogeneity and then were degassed in a vacuum chamber. The mixed, uncured SSA was coated onto a polytetrafluoroethylene (PTFE) film at a thickness of 0.9 mm, and placed in a forced air oven at a temperature of 130°C for 4 min to cure the SSA. The cured laminate was removed from the oven and allowed to cool to ambient temperature. A second PTFE film was applied using a 6.8 kg (15 lb.) rubber coated roller to ensure complete and consistent contact between SSA and PTFE. The film was allowed to rest for 24 h after which a disc was cut from the SSA laminate using a 24 mm stainless steel punch. Dynamic frequency sweeps (1–100 rad/s) were conducted on SSA with a TA ARES-G2 rheometer at 32°C using 25 mm stainless steel parallel plates and a gap of 0.5 mm with a 10% strain (in the linear viscoelastic region). Data collection was set for 5 pts./decade.

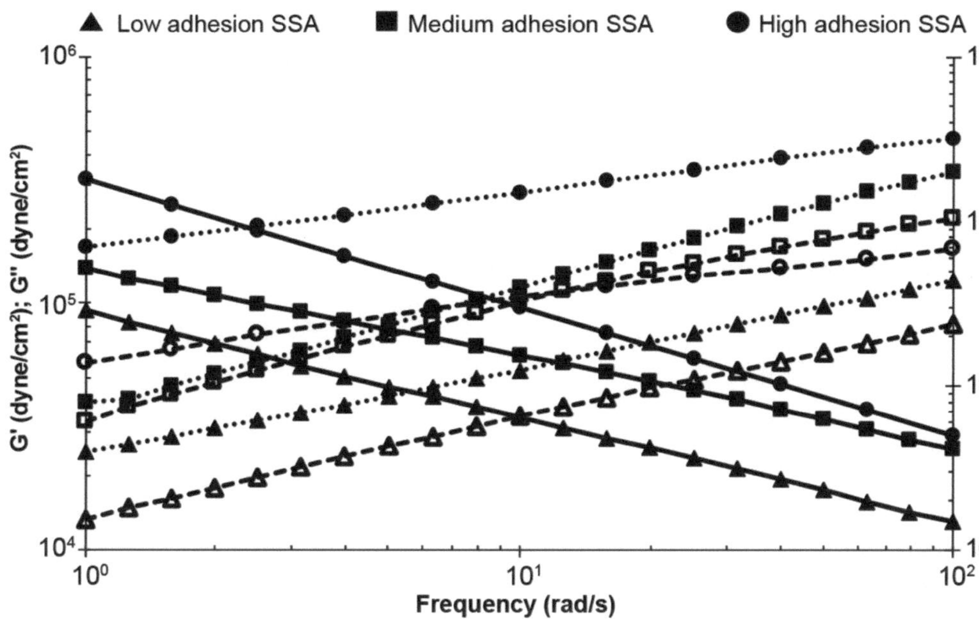

Figure 13. Frequency sweep of three typical SSA.

The frequency sweep test is the most suitable rheological test to assess SSA adhesive properties in the final application. The viscoelastic behavior at low frequencies is related to the bonding step which occurs at low deformation rates and is linked to the SSA ability to wet the surface. Alternatively, the viscoelastic behavior at high frequencies is related to debonding (peel) which occurs at high deformation rates and is linked to the elasticity and energy dissipation during the removal. SSAs with varying adhesive levels can be effectively discriminated based on their rheological profiles. The rheological characterization agrees with the results experienced by skin adhesion, where adhesives with higher G′ and G″ provide higher skin adhesion.

This rheology methodology should be an effective tool and a suitable starting point to understand the structure-property relationships of the SSA technology. It should also provide a means to separate the innate adhesive performance from the influences of substrates. Understanding the relationships between the SSA chemistry, adhesion and rheological profiles will provide key and essential information on structure-property relationships to push the boundaries of SSA even further.

11. Conclusion

Silicone adhesives have been safely and effectively used in a variety of medical applications and are notably present in drug delivery and wound care applications because of the unique benefits and properties provided. Continued investigation has resulted in recent, innovative product developments using established silicone adhesive technologies including innovative

TDDS designs, wound care devices that prevent scar formation and those that are loaded with antimicrobial actives. Adhesive chemistry research has resulted in novel chemistries that combine seemingly incompatible acrylate and silicone adhesive technologies, whereas advances in measurement techniques have brought about clearer understanding of adhesive structure property relationships, avoiding many pitfalls experienced by previous researchers. Despite being used for several decades, the number and variety of recent developments suggest that identifying new medical applications of silicone adhesives remains relevant and the extent to which it may be used has not yet been tapped.

Acknowledgements

The authors wish to acknowledge the following without whom this work would not have been possible; Chana Evans, Dave Gantner, Roger Gibas, Tim Mitchell, and Audrey Wipret. We would also like to acknowledge the work of Dr. Meng Gu and the microscopy team at The Dow Chemical Analytical Department for their assistance and collaboration.

Author details

Gerald K. Schalau II[1]*, Alexis Bobenrieth[2], Robert O. Huber[1], Linda S. Nartker[1] and Xavier Thomas[3]

*Address all correspondence to: g.k.schalau@dowcorning.com

1 The Dow Chemical Company, Food, Pharma and Medical, Midland MI, USA

2 The Dow Chemical Company, Food, Pharma and Medical, Seneffe, Belgium

3 The Dow Chemical Company, Food, Pharma and Medical, Saint-Denis, Paris, France

References

[1] Robb WL. Thin silicone membranes - their permeation properties and some applications. Annals of the New York Academy of Sciences. 1968;**146**:119

[2] Colas A, Curtis J. Silicone biomaterials: History and chemistry. In: Ratner BD, Hoffman AS, Schoen FJ, Lemons JE, editors. Medical Applications of Silicones in Biomaterials Science. 2nd ed. Elsevier Inc; 2004. pp. 80-86

[3] Noll W. Chemistry and Technology of Silicones. 2nd ed. New York, New York: Academic Press; 1968. pp. 1-61

[4] Schalau G, Aliyar H. Silicone excipients in pharmaceutical drug delivery applications. In: Narang A, Boddu S, editors. Excipient Applications in Formulation Design and Drug Delivery. 1st ed. Switzerland: Springer International; 2015. pp. 423-462

[5] Owens MJ. Surface chemistry and applications. In: Clarson SJ, Semlyen JA, editors. Siloxane Polymers. 1st ed. Englewood Cliffs, CA: PTR Prentice Hall; 1993. pp. 312-314

[6] Lin S, Durfee LD, Ekeland RA, McVie J, Schalau GK II. Recent advances in silicone pressure sensitive adhesives. Journal of Adhesion Science and Technology. 2007;**21**:605-623

[7] Gurtner GC, Dauskardt RH, Longaker MT, Yock P. Devices and Bandages for the Treatment and Prevention of Scars and/or Keloids and Methods and Kits thereof. US patent no. 7683234 B2; 2010

[8] Neodyne Biosicences Inc. Embrace [Internet]. 2017. Available from: http://embracescartherapy.com [Accessed: Sep 01 2017]

[9] Venkatraman S, Gale R. Skin adhesives and skin adhesion 1. Transdermal drug delivery systems. Biomaterials. 1998;**19**(13):1119-11136

[10] Pastore MN, Kalia YK, Horstman M, Roberts MS. Transdermal patches: History, development and pharmacology. British Journal of Pharmacology. 2015;**172**(9):2179-2209

[11] Woodard JT, Metevia VL. Transdermal Drug Delivery Devices with Amine-Resistant Silicone Adhesives. US patent no. 4655767. 1987

[12] The International Pharmaceutical Excipients Council of the Americas. Inactive Ingredient Database Issues with ANDAs Backgrounder Document [Internet]. 09-12-2011 . Available from: https://www.fda.gov/downloads/drugs/developmentapprovalprocess/howdrugsaredevelopedandapproved/approvalapplications/abbreviatednewdrugapplicationandagenerics/ucm291010.pdf [Accessed: 01-09-2017]

[13] Aliyar HA, Schalau GK II. Recent developments in silicones for topical and transdermal drug delivery. Therapeutic Delivery. 2015;**6**(7):827-839

[14] Yeoh T. Profiles of recently approved transdermal drug delivery systems (TDDS) Part 2: matrix-type fentanyl transdermal systems- design attributes. Transdermal. 2011;September:9-15

[15] United State Food Drug Administration. Guidance for Industry, Residual Drug in Transdermal Drug Delivery Systems; August 2011

[16] Miranda J, Sablotsky S. Solubility Parameter Based Transdermal Drug Delivery System and Method for Altering Drug Saturation Concentration. Canada patent application no. 2110914 A1. 1993

[17] Schalau GK II, Huber RO, Nartker LS, Thomas X. Novel silicone based adhesive technology for transdermal therapy systems. Therapeutic Delivery. 2017;**8**(4):175-178

[18] Sachdeva S, Goswami T, Audett J. Transdermal Delivery System. International Patent Application no. PCT US2016050904. 2017

[19] Noven Pharmaceuticals. Innovations in Passive Transdermal Drug Delivery: High Doses in a Small Patch. In: 33rd Annual Meeting of the Controlled Release Society; July 23; Vienna Austria; 2006

[20] Kanios D, Nartker L, Mitchell T, Evans C, Menjoulet T. Blend/matrix compatibility utilizing silicone/acrylic hybrid pressure sensitive adhesives in transdermal drug delivery systems. In: American Association of Pharmaceutical Scientists Annual Meeting and Exposition; October 23-27; Washington D.C., USA; 2011. Poster T2237

[21] Loubert GL, Menjoulet TA, Mitchell TP, Thomas XJ. Silicone acrylate Hybrid Composition and Method of Making Same. US patent no. US 8614278 B2. 2013

[22] Liu Y, Paul CW, Ouyang J, Foreman PB, Sridhar LM, Shah S. Silicone Acrylic Hybrid Polymer-based Adhesives. US patent no. 8580891B2. 2013

[23] Kanios D, Nartker L, Mitchell T, Evans C, Menjoulet T, Huber R. The effect of silicone/acrylic hybrid pressure sensitive adhesive in controlling in-vitro permeation and delivery profile from transdermal drug delivery systems. In: American Association of Pharmaceutical Scientists Annual Meeting and Exposition; October 23-27; Washington D.C. USA; 2011. Poster T2238

[24] Evans, CW, Huber RO, Nartker LS, Schalau II GK, Thomas X, Toth S. Multi-phase silicone acrylic hybrid visco-elastic compositions and methods of making same. PCT Filing US2016/016674; 2016

[25] Sweet R, Ulman K. Integrating Rheological Tools into the Development and Characterization of Silicone Adhesives. Dow Corning Corporation white paper. 1997;Form no. 51-965A-97

[26] Ulman K, Sweet R. The Correlation of Tape Properties and Rheology. Dow Corning Corporation white paper. 1998;Form no. 51-979-01

[27] Class JB, Chu SG. The viscoelastic properties of rubber-resin blends. I. The effect of resin structure. Journal of Applied Polymer Science. 1985;**30**(2):805-814

[28] Class JB, Chu SG. The viscoelastic properties of rubber-resin blends. II. The effect of resin molecular weight. Journal of Applied Polymer Science. 1985;**30**(2):815-824

[29] Class JB, Chu SG. The viscoelastic properties of rubber-resin blends. III. The effect of resin concentration. Journal of Applied Polymer Science. 1985;**30**(2):825-842

[30] Yang H. Water based polymers as pressure sensitive adhesives - viscoelastic guidelines. Journal of Applied Polymer Science. 1995;**55**(4):645-652

[31] Wolff H-M, Irsan, Dodou K. Investigations on the viscoelastic performance of pressure sensitive adhesives in drug-in-adhesive type transdermal films. Pharmaceutical Research. 2014;**31**(8):2186-2202

[32] Chu SG. Viscoelastic properties of pressure sensitive adhesives. In: Satas D, editor. Handbook of Pressure Sensitive Adhesives. 1st ed. New York: Van Norstand Reinhold Co; 1989. pp. 158-203

[33] Chang E. Viscoelastic windows of pressure sensitive adhesives. The Journal of Adhesion. 1991;**34**(1-4):189-200

[34] Nartker L, Girboux A-L, Bobenrieth A, Thomas X, Huber R, Sivanand P. Rheology Approach to Assess the Effect of Formulating Skin Permeability Enhancers with Pressure Sensitive Silicone Adhesives. In: American Association of Pharmaceutical Scientists Annual Meeting and Exposition; October 25-29; Orlando FL, USA; 2015. Poster W5142

[35] Thomas X. Silicone Adhesives in Healthcare Applications. Dow Corning Corporation white paper; 2003;Form no. 52-1057-01

[36] Pocknell D. Surgical dressing. European patent no. EP 0322118 B1; 1992

[37] Grisoni BFR, Pocknell D. Organosiloxane Gel-Forming Compositions and use thereof. European patent no. EP 0322118 B1; 1992

[38] Platt AJ, Phipps A, Judkins K. A comparative study of silicone net dressing and paraffin gauze dressings in skin grafted sites. Burns. 1996;**22**(7):543-543

[39] Gantner DC, Schalau II GK, Thomas X. Soft skin Adhesive Gels and Liners: New Formulating Options for Tailored Solutions. Dow Corning Corporation; 2007;Form no. 52-1113-01

[40] Yang L, Ditizio V. Antimicrobial Silicone based Wound Dressings. PCT no. CA2011/000712; 2011

Chapter 7

Adhesives: Applications and Recent Advances

Elena Dinte and Bianca Sylvester

Additional information is available at the end of the chapter

http://dx.doi.org/10.5772/intechopen.71854

Abstract

Adhesives can be defined as social substances capable to join permanently to surfaces, by an adhesive process. This process involves two dissimilar bodies being held in intimate contact such that mechanical force or work can be transferred across the interface. Since their early discovery by the Egyptians—3300 years ago—intensive research efforts have been made with the purpose of obtaining high-quality, biocompatible adhesives. Bitumen, tree pitches and beeswax—used in ancient and mediaeval times—were replaced by rubber cements and natural and synthetic components; nowadays, the focus is being mostly on eco-friendly adhesives. Starting with a brief history of adhesive use, this chapter then proceeds to cover the main industrial, biomedical and pharmaceutical applications of adhesives. Additionally, we focus on the new generation of adhesives, based on modern technologies such as nanotechnology, derivatised polymers, and biomimetic adhesives. The limited raw materials and the negative impact of synthetic adhesives on both human health and environment impose that further research is conducted with regard to renewable materials, in order to obtain environmentally safe bioadhesives that best fit their applicability domains.

Keywords: industrial adhesives, bioadhesives, mucoadhesives, sealant, tissue

1. Introduction

Adhesives are social substances and can be defined as a mixture in a liquid or semi-liquid state, capable to join permanently to surfaces, by an adhesive process. The word 'adhesive' can be used either as a noun or as an adjective, defining substances which tend to adhere or stick to other substances. Adhesion refers to the interaction of the adhesive surface with the substrate surface, and it involves two dissimilar bodies being held in intimate contact such that mechanical force or work can be transferred across the interface. Several theories that explain the adhesion process have been postulated, the forces involved in the process being van der Waals forces, chemical bonding, or electrostatic attraction. On the other hand, the mechanical strength of the system depends not only on the interfacial forces but also on the mechanical properties of the interfacial area, as well as the two bulk phases [1, 2].

Adhesives were first mentioned in history 3300 years ago, when Egyptian carvings depicted the glueing of a piece of veneer to what appears to be plank of sycamore. In ancient and mediaeval times, bitumen, tree pitches and beeswax were used as sealants and adhesives [3]. In the nineteenth century, rubber cements were introduced, but decisive advances in adhesive technology awaited the twentieth century, when natural adhesives have been improved and many synthetics components have been developed. Adhesives are essential components of shoes, automobiles, cartons, furniture and non-woven fabrics and a host of other products. The aerospace field was the first sector that promoted the use of adhesives in the aircraft manufacturing process; hence, the growth of the aircraft and aerospace industries has influenced adhesive technology in a great extent. The requirement to get a high degree of structural strength and a high resistance to fatigue promoted the development and production of high-performance adhesive materials, which found various domestic and industrial applications [3–5].

The raw materials used as adhesives are mainly polymeric materials, both natural and synthetic. Taking into consideration the costs, natural products (such as starch, dextrin, casein, naturally gums) are still important; however, synthetic ones have largely taken over the adhesive industry, both as modifiers of natural materials and, more importantly, as high-strength, moisture-resistant additives capable of being produced in many readily usable forms.

Among the key factors influencing the evolution of adhesives are globalisation, the maturity of technological processes and governmental regulations worldwide, militating for the usage of non-volatile adhesives, including epoxies, cyanoacrylates and urethanes, to the detriment of solvent-based adhesives [6].

Adhesives are also used in the healthcare sector, thus having broad applications in dentistry, medical and pharmaceutical field. Various modern adhesives are used in medicine and dentistry, in direct physiological interactive modes or to assemble thousands of medical devices. In the pharmaceutical field, the use of adhesives aims to design modern pharmaceutical systems, in order to optimise drug release rate, as well as targeted drug delivery. This favours a more efficient use of the pharmacological potential of the active substance, leading to an increased treatment efficiency, with the reduction of overall dosages and hence of the adverse reactions [7–9].

Nowadays, the focus is not only on the production of high-quality adhesives using modern technologies such as nanotechnology but also on the production of eco-friendly adhesives, named 'green' adhesives, for all domains of applicability [6, 10–12].

2. Adhesives for industrial applications

Adhesives are designed for specific applications. Besides their role in the adhesion process, they can be used for other purposes, such as sealing agents, in order to eliminate the effect of self-loosening caused by dynamic loads, sealing of areas to prevent oxidation and corrosion, waterproofing, etc. Sealants can be used as electrical or thermal insulators, fire barriers and products for smoothing, filleting or flying. The materials that are used as sealants have lower

http://dx.doi.org/10.5772/intechopen.71854

strength than those used as adhesives, because sealant formulations contain large amounts of inert filler material for cost reduction and gap filling purposes. Certain sealants, like adhesives, can be used to assemble parts, and many adhesives can be used to seal. The adhesives and sealants are mainly used to bond the following substrates: metals, plastics (thermosets and thermoplastics), composites, foams, elastomers, wood and wood products, glass and ceramics and sandwich and honeycomb structures [1–3, 5, 6].

The main areas using industrial adhesives are the following:

1. Construction: floor tile and continuous flooring installation, ceramic tile installation, countertop lamination, manufacture of prefabricated beams and trusses, carpet adhesives, flooring underlayment adhesives, installation of prefinished panels, joint cements, drywall lamination adhesives and covering installations.
2. Consumer adhesives: model and hobby supplies, decorative films, school and stationery products.
3. Packaging: carton-side seam and closures, composite bonding of disposable products, bags, labels, cups, cigarette and filter manufacture, speciality packages (cosmetics, toiletries), composite containers and tubes.
4. Tapes: packaging, industrial, surgical, masking, and consumer tapes.
5. Transportation: auto, truck and bus assemblies, weatherstrip and gasket bonding, aircraft and aerospace structural assemblies.
6. Other rigid bondings: shake-proof fastening; furniture manufacture; manufacture of millwork, doors, kitchen cabinets and vanitories; appliance assembly and trim attachment; TV, radio and electronics assembly and machinery manufacture and assembly.
7. Other non-rigid bondings: apparel laminates, shoe assembly, sports equipment, book binding, rug backing, flock cements, air and liquid filter manufacture, etc. [1].

The main adhesives for industrial applications are summarised in **Table 1**.

Adhesive type	Applications	Refs.
Natural		
Animal source: albumin, animal glue, casein, shellac, beeswax	Wood industry Food industry	[13]
Vegetable source: Natural resins: gum arabic, tragacanth, colophony	Paper, paperboard, light wood, cork	[14]
Oils and waxes: carnauba wax, different oils	Sealants in wood and metal industry	[15]
Proteins: soybean Carbohydrates: starch, dextrin	Wood industry Paper industry	[10, 15]
Mineral: Inorganic minerals: silicates, magnesia, phosphates, sulphur Mineral waxes: paraffin Mineral resins: amber	Wood industry	[1]

Adhesive type	Applications	Refs.
Bitumen: asphalt	Binder in roads, roofing and flooring, installation of asphalt tile, sealant	[16]
Synthetic		
Elastomers: Natural rubber and derivatives	Bonding paper, plastic, leather, shoe industry	[1]
Synthetic rubber: butyl, polyisobutylene, polybutadiene blends, polyisoprenes, polychloroprene, polyurethane, silicone, polysulfide, polyolefins	Bonding rubber to itself and plastic materials or metals; forms good bonding with most plastic films, glass, wood Some of them are sealants	[17, 18]
Reclaimed rubber	Bonding paper, rubber, plastic and ceramic tile, adhesive for electrical installations	[1]
Neoprene rubber	Bonding weather stripping and fibrous soundproofing materials to metal Sealant	[1]
Nitrile rubber	Bonding plastic films to metals and fibrous materials and nylon to nylon and other materials Sealant	[19]
Thermoplastics:		
Cellulose derivatives: acetate, acetate-butyrate, caprate, nitrate, methyl cellulose, hydroxyl ethyl cellulose, ethyl cellulose, carboxymethyl cellulose	Bonding non-metallic material: wood, leather, plastic, paper Used in packaging	[10, 20]
Polyacrylates: methacrylate and acrylate polymers, cyanoacrylates	Generally, adhesives for metals, some of them for plastic; thread lockers, thread sealants, retaining compounds, gasket Electronics, toys, cosmetic packaging, automotive sealants Bonds only to non-porous materials Glass industry	[16, 21]
Anaerobic: based on synthetic resins (acrylates)	Secure, seal and retain machined, threated or similarly close-fitting parts	[22–24]
Polyesters (saturated): polystyrene, polyamides	Sealants for potting, moulding and encapsulating	[1]
Vinyl polymers and copolymers: polyvinyl acetate, polyvinyl alcohol, polyvinyl chloride, polyvinylidene chloride	Bond porous materials Paper, wood and general packaging applications	[1]
Thermosettings		
Amino plastics: urea and melamine formaldehydes	Wood industry	[15]
Epoxies: epoxy polyamide, epoxy bitumen, epoxy polysulfide, epoxy nylon	Ability to bond many substrates Sealants in constructions, electricity	[25]
Phenolic resins and derivatives: phenol and resorcinol, formaldehydes, phenolic nitrile, phenolic neoprene, phenolic epoxy	Wood industry Metal industry	[26, 27]

Table 1. Industrial applications of the main adhesives (adapted from ref. [1]).

3. Adhesives for biomedical and pharmaceutical applications

The use of adhesives in the medical field has been restricted for a certain while, to the production of self-adhesive strips or plasters. The first reported pressure-sensitive adhesive used in the composition of bandage materials was natural rubber, followed by synthetic rubber, and, lastly, polyacrylic acid ester-based adhesives gained significance nowadays. Various medical adhesive tapes/dressings/devices are used to cover and protect wounds, to seal the skin edges of a wound or to support an injured part of the body [28, 29]. Advanced adhesives have a wide range of biomedical and pharmaceutical applications and are currently used in various medical procedures, as medical devices: restorative dental filings, blood transfusions, anaesthetic administration, intravenous drug delivery, heart bypass surgery, urological surgery and plastic surgery (**Table 2**) [7, 30–33].

Adhesive type	Applications	Refs.
Natural or biological		
Albumin	Haemostat in vascular and cardiac surgeries	[64, 76]
Alginate	Binds the tissue, even after exposure to an aqueous environment Adhesive, pharmaceutical excipient: stabilising agent, tablet and capsule disintegrant, tablet binder	[76, 77]
Chitosan-based adhesives	Bioadhesives, antibacterial, haemostatics, wound-closing agents Pharmaceutical excipient: bioadhesives coating agents, disintegrants, film-forming agents, tablet binders	[64, 76, 78]
Collagen-based adhesives	Haemostatics for general and vascular surgeries, retroperitoneal injuries, sealants (clot formation), wound dressing	[64]
Chondroitin sulphate glue	Bonds the native cartilage tissue, wound-closing agent Bonds implants, seals corneal incisions	[67, 79]
Fibrin-based adhesives	Haemostatics, sealants (clot formation), wound-closing agents	[32, 80]
Gelatine and gelatine-based products	Haemostatics in various surgical procedures and anatomical sites (clot formation), sealants, adhesives	[64, 65]
Hyaluronate sodium	Bioadhesive, sealant Plastical surgery Pharmaceutical excipient: humectant, lubricant, matrix for sustained release	[81–83]
Synthetic and semisynthetic		
Cyanoacrylates	Tissue adhesives in surgical procedures: vascular surgery, urology, cosmetic surgery, skin graft fixation Fistula repair, closure of hernia incisions Endoscopic, laparoscopic and interventional radiology procedures Sealants in dentistry Disposable plastic medical devices Transdermal patches	[33, 69, 71, 84]

Adhesive type	Applications	Refs.
Dendrimers	Studied as sealant agents in corneal incisions	[28, 67]
Polymers and their hydrogels		
PEG-based adhesives	Sealants in gynaecologic and colorectal procedures Sealing of fluid leaks, acute aortic dissection, haemostasis in anastomotic bleeding wound closure Bone fixation, sealant for vascular graft, haemostatic, wound-closing agent	[67, 75, 79]
Carbomer	Bioadhesive Pharmaceutical excipient: release-modifying agent, tablet binder, viscosity-increasing agent, suspending agent, tablet binder	[37, 58]
Polycarbophil	Bioadhesive Pharmaceutical excipient: absorbent, controlled-release tablet binder, emulsifying agent, thickening agent, suspending agent	[58, 85]
Polyethylene oxides and derivatives	Bioadhesive, surgical tissue adhesive Pharmaceutical excipient: tablet binder, thickening agent	[58, 86]
Poly(methyl vinyl ether/maleic anhydride)	Bioadhesive Pharmaceutical excipient: colour dispersant, complexing agent, film former, emulsion stabiliser, viscosity-increasing agent	[58]
Povidone	Bioadhesive Pharmaceutical excipient: disintegrant, tablet binder	[58]
Urethane-based	Sealant, tissue adhesive Fixation of vascular graft and bone Cosmetic surgery Preventing seroma formation in abdominoplasty	[67, 87]
Nano-enabled adhesive materials		
Polar lipids: glyceryl monooleate	Bioadhesive, pharmaceutical excipient Nano-vehicle for many active substances	[60–62]
Nanoparticles (based on different components)	Bioadhesives Antibacterial properties Wound-closing agents	[64, 87, 88]
Biomimetic adhesives		
Marine mussel extract adhesives	Bioadhesives, repairing gestational fatal membrane ruptures Islet transplantation at extrahepatic sites Prosthetic mesh fixation Some derivatives: haemostatic and wound-closing agents	[31, 33, 89]
Gecko-inspired adhesives	Wound sealing, suture and staple replacement Suture/staple replacement/supplements Waterproof sealant for hollow organ anastomoses and for prevention of air leaks in lung resection Haemostatic, wound dressing, mesh grafts (ulcers, hernias, burns)	[33, 64]

Adapted from [28, 58, 64, 67, 76]

Table 2. Summary of the representative adhesives with biomedical and pharmaceutical applications.

When an adhesive comes in contact with a biological tissue, it is named 'bioadhesive'. Bioadhesion is the capacity of a compound to adhere to the biologic substrate for a long time. When the biologic substrate is represented by a mucosa, the phenomenon is called mucoadhesion. The mucoadhesive materials interact with the glycoproteins in the mucus covering the epithelia of the mucosae. The generally accepted idea is that the mucoadhesion process involves several stages: wetting and swelling of the hydrophilic polymer which allows its contact with the biologic tissue and the interpenetration of the polymer chain with the molecules in the mucin, which leads to an adequate interpenetration of the substrate and creates a semi-permanent adhesive bonding. Other theories explain the forces that underpin bioadhesion: van der Waals forces, hydrogen bondings, disulphide bridges, hydration forces, hydrophobic interactions, steric forces, covalent bonds, etc. [34, 35].

Bioadhesive drug delivery systems are promising systems for delivery of numerous and various active substances, from common ones to prebiotics [36], herbal products [37] and proteins [38]. They present themselves as solids, semisolids and liquids (gels, films, tablets, etc.) in conventional formulations or as nanoparticulated systems, designed for various routes like oral route [39, 40], skin route [41, 42] or mucosal route (buccal [43], ocular [44], vaginal [45], nasal [46], oesophageal [47]). The use of nanoparticulate bioadhesive systems can substantially improve the absorption of the active substances while offering protection against certain factors. Moreover, nanoparticulate bioadhesive systems represent potential targeted protein delivery systems [48].

Testing of the mucoadhesion represents a very important objective during the development of bioadhesive systems for drug release, as it can reveal the compatibility with the other components of the system, their stability and the strength of the adhesion capacity. Most methods used for the control of the mucoadhesive capacity of drug systems cited in the literature are based on the measurement of the force required to the disruption of the adhesive bond between the model membrane and the adhesive material [49].

The main excipients of a bioadhesive system are represented by bioadhesive polymers and their structural and functional characteristics having a decisive influence on the bioadhesion. These polymers are hydrophilic matrices consisting of a reticular network that swell in contact with water without being dissolved. The polymeric materials are available in a wide variety of molecular weights and compositions which adsorb the water, swell and generate a gel structure (hydrogel). These swollen gels act as a reservoir and ensure a prolonged release of the active substances dispersed in the meshes of the polymeric network. After swelling, a relaxation of the network chains takes place, and the incorporated drug substances are released through the spaces inside the network. The property to control the release rate of the biologically active substance, the bioadhesive performance and the nontoxic profile of the polymer are primordial criteria for the selection of a bioadhesive agent used in the designing of controlled drug release systems. The hydrogel-forming polymers used for the preparation of mucoadhesive systems are represented mainly by polyacrylates (Carbopol and polycarbophil), polyethylene oxide, polyvinyl alcohol, poly(*N*-acryloylpyrrolidine), reticulated gelatin, sodium alginate, natural gums (guar, xanthan, karaya), cellulose ethers, etc. [34, 35]. The polymer only plays the role of a vector or a host matrix, and it should have as neutral behaviour as possible towards the active drug substance. The polymer-drug system must be chosen in such a way as to avoid chemical reactions between components, which may lead to the degradation of the active substance. The existence of possible interactions between the host polymer matrix and the drug can be tested at an atomic and molecular scale through well-known techniques such as differential scanning calorimetry (DSC) and spectroscopic methods such as Fourier transform infrared spectroscopy (FTIR), X-ray diffraction, electron spin resonance (ESR) spectroscopy, nuclear magnetic resonance (NMR) spectroscopy, Raman spectroscopy and others [50–52].

Different hydrogels can be combined together or with other components, obtaining derivatised polymers, with the purpose of improving their jellifying and adhesion properties [42, 53–55]. An innovative approach in the synthesis of hydrogels aims at the improvement of their mechanic properties. This can be achieved by grafting various functional groupings that allow self-assembling in aggregates or triblocks polymers (Poloxamers) that can change their consistency in the presence of certain stimuli (pH, temperature, etc.). Heat-sensitive polymers are of particular interest as drug vehicles, as they allow the preparation of formulations with a lower consistency and are easy to apply and develop a good contact in the presence of physiological medium, by forming in situ gels [47, 56, 57].

Besides their role as adhesives, bioadhesive polymers can also fulfil other functions in a pharmaceutical system: diluent, disintegrant, viscosity enhancer, etc. Their nature, quantity and association with other excipients highly influence the final drug product quality [58].

Bioadhesive polymers are used in many medical devices and drug delivery systems, including transdermal patches. Transdermal patch technology is another application of biomedical adhesives, ensuring drug delivery to the bloodstream through the skin; this is a highly effective method of drug administration, as the drug is incorporated into a membrane (made of adhesive) that sticks the patch to the skin and controls the rate at which the drug is absorbed. These systems ensure that the drug is continuously administered throughout the day, avoiding the fluctuations of the plasmatic concentration, usually associated with orally administered drugs. The applications of patches include hormone replacement therapy, pain cessation, smoke cessation and treatment of various cardiovascular pathologies [30, 59].

An alternative for polymeric materials with bioadhesive properties is polar lipids (e.g. glyceryl monooleate). Polar lipids are water-insoluble amphiphilic molecules that swell when in contact with water, associate and form various types of aggregates (spheric, hexagonal micelle, lamellar phase, cubic phase). The cubic crystalline phase has the aspect of a transparent, rigid, viscous gel with good mucoadhesive characteristics and can incorporate hydrophilic, amphiphilic as well as lipophilic drug substances. These bioadhesive properties make it an in situ forming biodegradable matrix, thus a potential nanostructured vehicle for prolonged drug release. Cubosomes represent a dispersion of the cubic crystalline phase (similar to liposomes which represent the dispersed lamellar crystalline phase). They are submicronic particles and present unique properties as they can incorporate drug substances with various polarities and molecular masses (rifampicin, carbamazepine, griseofulvin, coenzyme Q10, lycopene, phytosterols, sodium diclofenac, etc.). Some studies revealed that cubic-phase dispersions can maintain high plasma oligopeptide levels for several hours and favour insulin absorption through the nasal mucosa in rats. Moreover, cubosomes lower the rate of enzymatic degradation of the oligopeptides (insulin, somatostatin) and enzymes. Multiple possibilities of using cubosomes as drug vehicles for various routes of administration (oral, parenteral, cutaneous or mucosae: nasal, ophthalmic, vaginal, buccal, periodontal pocket) have been reported so far in literature [60–63].

Wound closure is a key step in the success of surgical procedures. Surgical adhesives represent a convenient method for wound closing, having several advantages: less pain, no suture removal, excellent cosmetic result, and localized drug release [64]. A tissue adhesive can be defined as any substance with characteristics that allow in situ polymerisation to cause adherence of tissue to tissue or tissue to non-tissue surfaces. Their applications include prostheses, bleeding control (haemostats) and serving as a barrier to gas and liquids (sealants) [65]. Many tissue adhesives and haemostats have been developed over the past 30 years, based on various materials. The existing models are not sufficient for an in-depth study of the attachment of adhesives to living tissues; therefore, the strength of the adhesive joint is determined experimentally, according to standard tests and under conditions that are close to a real surgical situation. A bioadhesive must exhibit distinct characteristics, depending on the targeted tissue [66]. The successful usage of a tissue adhesive depends on its specific indications/limitations, which need to be thoroughly analysed by surgeons in order to choose the best product [67].

There are different commercially available tissue adhesives: natural or biological, synthetic, semisynthetic and biomimetic. The biomimetic action of adhesives is based on algae's abilities to wet a surface and the gecko's ability to 'adhere' on to surfaces. Surgical adhesives and sealants based on natural polymers offer a more biocompatible alternative to synthetic glues. The composition of a new bioadhesive material from fish parasite *Neobenedenia girellae* was recently reported, proteomic analysis revealing that the adhesive is mainly composed of cytoskeletal proteins such as actin, keratin and tubulin [28, 68].

Dental adhesives are intended to provide retention to composite fillings or composite cements. They can be defined as solutions of resin monomers (with both hydrophilic and hydrophobic groups) that make the resin-dental substrate interaction possible. Some of them are also used for their protective effect against enamel erosion. The failure of restorations occurs more often due to the inadequate sealing, with subsequent discoloration of the cavity margins, rather than loss of retention. Recent development of dental adhesives has greatly simplified the application procedure, as opposed to classical bonding agents (multistep systems), with the purpose of reducing technique sensitivity and manipulation time [69–74].

While older generation of mucoadhesive polymers lacks specificity and targeting capability, newer polymers—falling into the second generation of mucoadhesives—can form covalent bonds with the mucus and the underlying cell layers, thus exhibiting improved chemical interactions and offering new possibilities for more specific drug-receptor interactions. Examples of such new-generation adhesives are thiolated and lectin-mediated mucoadhesive polymers [75].

4. Conclusions and perspectives

Adhesives have been used in industry for decades; however, the environmental influence of adhesives has not been investigated up until recently. Therefore, one must emphasise on the necessity to develop steps that will allow obtaining environmentally safe and high-quality adhesives that best fit their applicability domains. Due to the limited raw materials (reserve of oil) and the negative impact of synthetic compounds on both human health and environment, natural and renewable resources represent an attractive alternative for the production of adhesives. One innovative alternative for synthetic reactive glues like cyanoacrylates is represented by gecko-inspired and marine-inspired bioadhesives (mussel proteins) or their combination; however, further research needs to be conducted in this direction. Taking into consideration the rapid development of bioadhesive market, more work needs to be done in both environmental and economic aspect.

Author details

Elena Dinte* and Bianca Sylvester

*Address all correspondence to: edinte@gmail.com

Department of Pharmaceutical Technology and Biopharmaceutics, University of Medicine and Pharmacy Iuliu Hațieganu, Cluj-Napoca, Romania

References

[1] Petrie EM. An Introduction to Adhesive and Sealants. In: Handbook of Adhesives and Sealants. First ed. McGraw-Hill Professional, New York, NY, USA; 1999. p. 2-48

[2] Ebnesajjad S. Adhesive Technology Handbook. 2nd ed. William Andrew Inc, Norwich, NY, USA; 2008

[3] Sunday OO. Strength of adhesive bonded joints: Comparative strength of adhesives. International Journal of Engineering and Technical Research. 2015;**3**(8):58-62 ISSN:2321-0869(O) 2454-4698 (P)

[4] Skeist I, Miron J. Introduction to adhesives. In: Skeist I, editor. Handbook of Adhesives. Boston: Springer; 1990

[5] Frihart CR. Wood adhesion and adhesives. In: Rowell RM, editor. Handbook of Wood Chemistry and Wood Composites. Boca Raton: CRC Press LLC; 2005. pp. 215-278

[6] Patel JP, Xiang ZG, Hsu SL, Schoch AB, Carleen SA, Matsumoto D. Characterization of the crosslinking reaction in high performance adhesives. International Journal of Adhesion and Adhesives. 2017;**78**:256-262

[7] Hoffman A. Hydrogels for biomedical applications. Advanced Drug Delivery Reviews. 2002;**43**:3-12

[8] Kashyap N, Kumar N, Ravi Kumar MNV. Hydrogels for pharmaceutical and biomedical applications. Critical Reviews in Therapeutic Drug Carrier Systems. 2005;**22**:107-149

[9] Vernengo AJ. Adhesive materials for biomedical applications. In: Adhesives–Aplications and Properties. Rijeka: InTechOpen; 2016. pp. 100-136

[10] Norstrom E, Fogelstrom L, Nordqvist P, Khabbaz F, Malmstrom E. Xylan–a green binder for wood adhesive. European Polymer Journal. 2015;**67**:483-493

[11] Song YH, Seo JH, Choi YS, Kim DH, Choi B-H, Cha HJ. Mussel adhesive protein as an environmentally-friendly harmless wood furniture adhesive. International Journal of Adhesion and Adhesives. 2016;**70**:260-264

[12] Santoni I, Pizzo B. Evaluation of alternative vegetable proteins as wood adhesives. Industrial Crops and Products. 2013;**45**:148-154

[13] Shellac. Available from: https://en.wikipedia.org/wiki/Shellac (accessed on 08-09-2017)

[14] Paiva D, Goncalves G, Vale I, Bastos MMSM, Magalhaes FD. Oxidized xanthan gum and chitosan as natural adhesives for cork. Polymer. 2016;**8**(259):1-13

[15] Cheng HN, Ford C, Dowd MK, He Z. Use of additives to enhance the properties of cottonseed protein as wood adhesives. International Journal of Adhesion and Adhesives. 2016;**68**:156-160

[16] Asphalt. Available from: https://en.wikipedia.org/wiki/Asphalt (accessed on 06-09-2017)

[17] Blyberg L, Serrano E, Enquist B, Sterley M. Adhesive joints for structural timber/glass applications: Experimental testing and evaluation methods. International Journal of Adhesion and Adhesives. 2012;**35**:76-87

[18] Moghadam PN, Yarmohamadi M, Hasanzadeh R, Nuri S. Preparation of polyurethane wood adhesives by polyols formulated with polyester polyols based on castor oil. International Journal of Adhesion and Adhesives. 2016;**68**:273-282

[19] Nitrile rubber. Available from: https://en.wikipedia.org/wiki/Nitrile_rubber (accessed on 06-09-2017)

[20] Farhat W, Venditti R, Quick A, Taha M, Mignard N. Hemicellulose extraction and characterization for applications in paper coatings and adhesives. Industrial Crops and Products. 2017;**107**:370-377

[21] Bucek A, Brablec A, Kovacik D, Stahel P, Cernak M. Glass bond adhesive strength improvement by DCSBD atmospheric-pressure plasma treatment. International Journal of Adhesion and Adhesives. 2017;**78**:1-3

[22] ThreeBond Group. About anaerobic adhesives, Technical information. Available from: https://www.threebond.co.jp/en/technical/seminar/adhesion2.html#no03 (accessed on 06-09-2017)

[23] Henkel. Loctite anaerobic adhesives. Available from: http://na.henkel-adhesives.com/industrial/anaerobic-adhesive-14883.htm (accessed on 06-09-2017)

[24] Ireland AJ, Sherriff M. Transition metal salt solutions and anaerobic adhesives in dental bonding. Dental Materials. 1999;**15**:243-249

[25] Heo JH, Lee JW, Lee B, Cho HH, Lim B, Lee JH. Chemical effects of organo-silanized SiO_2 nanofillers on epoxy adhesives. Journal of Industrial and Engineering Chemistry. 2017;**54**:184-189

[26] Yelle DJ, Ralph J. Characterizing phenol-formaldehyde adhesive cure chemistry within the wood cell wall. International Journal of Adhesion and Adhesives. 2016;**70**:26-36

[27] Phenolic Resin Adhesives. Available from: http://polymerdatabase.com/Adhesives/Phenolic%20Adhesive.html (accessed on 06-09-2017)

[28] Bouten PJM, Zonjee M, Bender J, Yauw STK, van Goor H, van Hest JCM, Hoogenboom R. The chemistry of tissue adhesive materials. Progress in Polymer Science. 2014;**39**:1375-1405

[29] Lund C. Medical adhesives in the NICU. Newborn infant. Nursing Review. 2014;**14**:160-165

[30] El-Gendy NA, Sabry NA, El-Attar M, Omar E, Mahmoud M. Transdermal delivery of salbutamol sulphate: Formulation and evaluation. Pharmaceutical Development and Technology. 2009;**14**(2):216-225

[31] Kim HJ, Hwang BH, Lim S, Choi B-h, Kang SH, Choi HJ. Mussel adhesion-employed water-immiscible fluid bioadhesive for urinary fistula sealing. Biomaterials. 2015;**72**:104-111

[32] Plat VD, Bootsma BT, van der Wielen N, Straatman J, Schoonmade LJ, van der Peet DL, Daams F. The role of tissue adhesives in esophageal surgery, a systematic review of literature. International Journal of Surgery. 2017;**40**:163-168

[33] Mehdizadeh M, Weng H, Gyawali D, Tang L, Yang J. Injectable citrate-based mussel-inspired tissue bioadhesives with high wet strength for sutureless wound closure. Biomaterials. 2012;**33**:7972-7983

[34] Andrews GP, Laverty TP, Jones DS. Mucoadhesive polymeric platforms for controlled drug delivery. European Journal of Pharmaceutics and Biopharmaceutics. 2009;**71**:505-518

[35] Kopecek J. Hydrogel biomaterials: A smart future? Biomaterials. 2007;**28**:5185-5192

[36] Pliszczak D, Bourgeois S, Bordes C, Valour JP, Mazoyer MA, Orecchioni AM, Nakache E, Lanteri P. Improvement of an encapsulation process for the preparation of pro- and prebiotics-loaded bioadhesive microparticles by using experimental design. European Journal of Pharmaceutical Sciences. 2011;**44**:83-92

[37] Patel NA, Patel M, Patel RP. Formulation and evaluation of polyherbal gel for wound healing. International Research Journal of Pharmacy. 2011;**1**(1):1-6

[38] Morishita M, Barichello JM, Takayama K, Chiba Y, Tokiwa S, Nagai T. Pluronic F-127 gels incorporating highly purified unsaturated fatty acids for buccal delivery of insulin. International Journal of Pharmaceutics. 2001;**212**:289-293

[39] Zhang X, Sun M, Zheng A, Cao D, Bi Y, Sun J. Preparation and characterization of insulin-loaded bioadhesive PLGA nanoparticles for oral administration. European Journal of Pharmaceutical Sciences. 2012;**45**:632-638

[40] Laulicht B, Cheifetz P, Tripathi A, Mathiowitz E. Are *in vivo* gastric bioadhesive forces accurately reflected by in vitro experiments? Journal of Controlled Release. 2009;**134**:103-110

[41] Borghi-Pangoni FB, Junqueira MV, de Souza Ferreira SB, Silva LL, Rabello BR, de Castro LV, Baesso ML, Diniz A, Caetano W. Preparation and characterization of bioadhesive system containing hypericin for local photodynamic therapy. Photodiagnosis and Photodynamic Therapy. 2017;**19**:284-297

[42] Garcia MC, Aldana AA, Tartara LI, Alovero F, Strumia MC, Manzo RH, Martinelli M, Jimenez-Karus AF. Bioadhesive and biocompatible films as wound dressing materials based on a novel dendronized chitosan loaded with ciprofloxacin. Carbohydrate Polymers. 2017;**175**:75-86

[43] Dinte E, Tomuta I, Iovanov RI, Leucuta SE. Design and formulation of buccal mucoadhesive preparation based on sorbitan monostearate oleogel. Farmácia. 2013;**61**(2): 284-297

[44] Calles JA, Tartara LI, Lopez-Garcia A, Diebold Y, Palma SD, Valles EM. Novel bioadhesive hyaluronan-itaconic acid crosslinked films for ocular therapy. International Journal of Pharmaceutics. 2013;**455**:48-56

[45] Bassi P, Kaur G. Polymeric films as a promising carrier for bioadhesive drug delivery: Development, characterization and optimization. Saudi Pharmaceutical Journal. 2017;**25**:32-43

[46] Jiao Y, Pang X, Liu M, Zhang B, Li L, Zhai G. Recent progress in bioadhesive microspheres via transmucosal administration. Colloids and Surfaces, B: Biointerfaces. 2016;**140**:361-372

[47] Mako A, Csoka G, Pasztor E, Marton S, Horvai G, Klebovich I. Formulation of thermoresponsive and bioadhesive gel for treatment of oesophageal pain and inflammation. European Journal of Pharmaceutics and Biopharmaceutics. 2009;**72**:260-265

[48] Agrawal P, Sing S, Singh RP, Sharma G, Mehata AK, Singh S, Rajesh CV, Pandey BL, Koch B, Muthu MS. Bioadhesives micelles of D-α-tocopherol polyethylene glycol succinate 1000: Synergism of chitosan and transferrin in targeted drug delivery. Colloids and Surfaces, B: Biointerfaces. 2017;**152**:277-288

[49] Mortazavi SA, Smart JD. An *in vitro* method for assessing the duration of mucoadhesion. Journal of Controlled Release. 1994;**31**:207-212

[50] Dinte E, Bodoki E, Leucuta SE, Iuga CA. Compatibility studies between drugs and excipients in the preformulation phase of buccal mucoadhesive systems. Farmácia. 2013;**61**(4):703-712

[51] Todica M, Pop CV, Dinte E, Farcau C, Astilean S. Preliminary investigation by Raman spectroscopy of some polymeric matrix with pharmaceutical applications. Modern Physics Letters B. 2007;**21**(16):987-995

[52] Todica M, Dinte E, Pop CV, Farcau C, Astilean S. Raman investigation of some polymeric gels of pharmaceutical interest. Journal of Optoelectronics and Advanced Materials. 2008;**10**(4):823-825

[53] Guo J, Wang W, Hu J, Xie D, Gerhard E, Nisic M, Shan D, Qian G, Zheng S, Yang J. Synthesis and characterization of anti-bacterial and anti-fungal citrate-based mussel-inspired bioadhesives. Biomaterials. 2016;**85**:204-217

[54] Bassi P, Kaur G. Bioadhesive vaginal drug delivery of nystatin using a derivatized polymer: Development and characterization. European Journal of Pharmaceutics and Biopharmaceutics. 2015;**96**:173-184

[55] Cheewatanakornkool K, Niratisai S, Sriamornsak P. Bioadhesiveness of thiolated pectin for buccal delivery of carbenoxolone sodium. Asian Journal of Pharmaceutical Sciences. 2016;**11**:124-125

[56] Mayol L, Quaglia F, Borzacchiello A, Ambrosio L, La Rotonda MI. A novel poloxamers/hyaluronic acid in situ forming hydrogel for drug delivery: Rheological, mucoadhesive and *in vitro* release properties. European Journal of Pharmaceutics and Biopharmaceutics. 2008;**70**:199-206

[57] Han I-K, Kim YB, Kang H-S, Sul D, Jung W-W, Cho HJ, Oh Y-K. Thermosensitive and mucoadhesive delivery systems of mucosal vaccines. Methods. 2006;**38**:106-111

[58] Rowe RC, Sheskey PJ, Quinn ME. Handbook of Pharmaceutical Excipients. 6th ed. Pharmaceutical Press. London, UK; 2009

[59] Donnely RF, McCarron PA, Zawislak AA, Woolfson AD. Design and physicochemical characterisation of a bioadhesive patch for dose-controlled topical delivery of imiquimod. International Journal of Pharmaceutics. 2006;**307**:318-325

[60] Liu Y, Zhang J, Gao Y, Zhu J. Preparation and evaluation of glyceryl monooleate-coated hollow-bioadhesive microspheres for gastroretentive drug delivery. International Journal of Pharmaceutics. 2011;**413**(1-2):103-109

[61] Nielsen LS, Helledi LS, Schubert L. Release kinetics of acyclovir from a suspension of acyclovir incorporated in a cubic phase delivery system. Drug Development and Industrial Pharmacy. 2001;**27**(10):1073-1081

[62] Spicer PT. Progress in liquid crystalline dispersions: Cubosomes. Current Opinion in Colloid & Interface Science. 2005;**10**:274-279

[63] Lopes LB, Lopes JL, Oliveira DC, Thomazini JA, Garcia MT, Fantini MC, Collett JH, Bentley MV. Liquid crystalline phases of monoolein and water for topical delivery of cyclosporin A: Characterization and study of *in vitro* and *in vivo* delivery. European Journal of Pharmaceutics and Biopharmaceutics. 2006;**63**(2):146-155

[64] Annabi N, Tamayol A, Shin SR, Ghaemmaghami AM, Peppas NA, Khademhosseini A. Surgical materials: Current challenges and nano-enabled solutions. Nano Today. 2014; **9**:574-589

[65] Pinkas O, Zilberman M. Novel gelatin-alginate surgical sealant loaded with hemostatic agents. International Journal of Polymeric Materials. 2017;**66**(8):378-387

[66] Marques DS, Santos JMC, Ferreira P, Correia TR, Correia IJ, Gil MH, Baptista CMSG. Photocurable bioadhesive based on lactic acid. Materials Science and Engineering: C. 2016;**58**:601-609

[67] Duarte AP, Coelho JF, Bordado JC, Cidade MT, Gil MH. Surgical adhesives: Systematic review of the main types and development forecast. Progress in Polymer Science. 2012;**37**:1031-1050

[68] Maffioli E, Nonnis S, Polo NC, Negri A, Forcella M, Fusi P, Galli P, Tedeschi GA. New bioadhesive material from fish parasite *Neobenedenia girellae*. Journal of Proteomics. 2014;**110**:1-6

[69] Sezinando A. Looking for the ideal adhesive–a review. Revista Portuguesa de Estomatologia, Medicina Dentária e Cirurgia Maxilofacial. 2014;**55**(4):194-206

[70] Da Silva Avila DM, Zanatta RF, Scaramucci T, Aoki IV, Torres CRG, Borges AB. Influence of bioadhesive polymers on the protective effect of fluoride aging erosion. Journal of Dentistry. 2017;**56**:45-52

[71] Sofan E, Sofan A, Palaia G, Tenore G, Romeo U, Migliau G. Classification review of dental adhesive systems: From the IV generation to the universal type. Annali Di Stomatologia (Roma). 2017;**8**(1):1-17

[72] Milia E, Cumbo E, Cardoso RJA, Gallina G. Current dental adhesives systems. A narrative review. Current Pharmaceutical Design. 2012;**18**:5542-5552

[73] Munoz MA, Luque I, Hass V, Reis A, Loguercio AD, Bombarda NHC. Immediate bonding properties of universal adhesives to dentine. Journal of Dentistry. 2013;**41**:404-411

[74] Marchesi G, Frasseto A, Mazzoni A, Apolonio F, Diolosa M, Cadenaro M, Di Lenarda R, Pashley DH, Tay F, Breschi L. Adhesive performance of a multi-mode adhesive system: 1-year *in vitro* study. Journal of Dentistry. 2014;**42**:603-612

[75] Agarwal S, Aggarwal S. Mucoadhesive polymeric platform for drug delivery: A comprehensive review. Current Drug Delivery. 2015;**12**(2):139-156

[76] Annabi N, Yue K, Tamayol A, Khademhosseini A. Elastic sealants for surgical applications. European Journal of Pharmaceutics and Biopharmaceutics. 2015;**95**:27-39

[77] De'Nobili Curto LM, Delfino JM, Soria M, Fissore EN, Rojas AM. Performance of alginate films for retention of L-(+)-ascorbic acid. International Journal of Pharmaceutics. 2013;**450**:95-103

[78] Szymanska E, Winnicka K. Preparation and *in vitro* evaluation of chitosan microgranules with clotrimazole. Acta Poloniae Pharmaceutica. Drug Research. 2012;**69**(5):509-513

[79] Suchaoin W, Bonengel S, Grießinger JA, de Sousa IP, Hussain S, Huck CW, Bernkop-Schnürch A. Novel bioadhesive polymers as intra-articular agents: Chondroitin sulphate-cysteine conjugates. European Journal of Pharmaceutics and Biopharmaceutics. 2016;**101**:25-32

[80] Kern N, Behrens AM, Srinivasan P, Rossi CT, Daristotle JL, Kofinas P, Sandler AD. Solution blow spun polymer: A novel preclinical surgical sealant for bowel anastomoses. Journal of Pediatric Surgery. 2017;**52**:1308-1312

[81] Anand S, Singisetti K, Srikanth KN, Bamforth C, Asumu T, Buch K. Effect of sodium hyaluronate on recovery after arthroscopic knee surgery. The Journal of Knee Surgery. 2016;**29**(6):502-509

[82] Salzillo R, Schiraldi C, Corsuto L, D'Agostino A, Filosa R, De Rosa M, La Gatta A. Optimization of hyaluronan-based eye drop formulations. Carbohydrate Polymers. 2016;**153**:275-283

[83] Hahn SK, EJ O, Miyamoto H, Shimobouji T. Sustained release formulation of erythropoietin using hyaluronic acid hydrogels crosslinked by Michael addition. International Journal of Pharmaceutics. 2006;**322**:44-51

[84] Angulo A, Sebastián I, Martínez FJ, Torregrosa R, Martín-Martínez JM, Madariaga AM. Comparative effectiveness of cyanoacrylate bioadhesives and monofilament suture in wound healing: A histopathological and physicochemical study in New Zealand white rabbit. Journal of Cytology & Histology. 2016;**7**:395

[85] Zhu Z, Zhai Y, Zhang N, Leng D, Ding P. The development of polycarbophil as a bioadhesive material in pharmacy. Asian Journal of Pharmaceutical Sciences. 2013;**8**:218-227

[86] Ghosh S, Cabral JD, Hanton LR, Moratti SC. Strong poly(ethylene oxide) based gel adhesives via oxime cross-linking. Acta Biomaterialia. 2016;**29**:206-214

[87] Melgar-Lesmes P, Morral-Luiz G, Solans C, Garcia-Celma MJ. Quantifying the bioadhesive properties of surface-modified polyurethane-urea nanoparticles in the vascular network. Colloids and Surfaces, B: Biointerfaces. 2014;**118**:280-288

[88] Yee W, Selvaduray G, Hawkins B. Characterization of silver nanoparticle-infused tissue adhesive for ophthalmic use. Journal of the Mechanical Behavior of Biomedical Materials. 2016;**55**:67-74

[89] Ai Y, Wei Y, Nie J, Yang D. Study on the synthesis and properties of mussel mimetic poly(ethylene glycol) bioadhesive. Journal of Photochemistry and Photobiology B: Biology. 2013;**120**:183-190